Julia Köller
Johann Köppel
Wolfgang Peters
Offshore Wind Energy

Julia Köller
Johann Köppel
Wolfgang Peters

(Editors)

Offshore Wind Energy

Research on Environmental Impacts

With 135 Figures

 Springer

DIPL.-ING. JULIA KÖLLER
E-mail:
koeller@ile.tu-berlin.de

PROF. DR. JOHANN KÖPPEL
E-mail:
koeppel@ile.tu-berlin.de

DR. WOLFGANG PETERS
E-mail:
peters@ile.tu-berlin.de

Berlin University of Technology

Institute of Landscape Architecture
and Environmental Planning,
Department for Landscape Planning
and Environmental Impact Assessment
Straße des 17. Juni 145
Secr. EB 5
10623 Berlin
Germany

Library of Congress Control Number: 2006929203

ISBN-10   3-540-34676-7   Springer Berlin Heidelberg New York
ISBN-13   978-3-540-34676-0   Springer Berlin Heidelberg New York

Springer is a part of Springer Science+Business Media
springer.com
© Springer-Verlag Berlin Heidelberg 2006

Cover design: Erich Kirchner
Typesetting: camera-ready by the editors
Production: Christine Adolph
Printing: Krips bv, Meppel
Binding: Stürtz AG, Würzburg

Printed on acid-free paper   30/3180/ca 5 4 3 2 1   SPIN 12061501

Foreword

Accompanying ecological research is an important prerequisite for the sustainable development of offshore wind power. After exaggerated fears of the possible environmental impacts of planned wind farms in the beginning, the results of the first phase of the research projects may contribute to a more differentiated and realistic assessment of the environmental impacts of offshore wind farms. The projects were initiated by the German Federal Ministry for the Environment, Nature Conservation and Nuclear Safety and its results are published in this book.

The research results published here as a synopsis provide an early contribution to a precautionary consideration of possible environmental impacts of the development of wind energy at sea, particularly in the exclusive economic zone. The coherent overall concept for the development of offshore wind energy pursued by the German federal government is remarkable and hitherto unique among large-scale domestic projects. The combination of standardised investigations and the Environmental Impact Assessment (EIA) of each wind farm together with basic examinations for subsequent monitoring during the progress of the wind farms' construction meant the early implementation of certain requirements. In the course of the introduction of the EU's Strategic Environmental Assessment (SEA) process these became mandatory only recently.

Thus, it became possible to plan the accompanying ecological research in greater depth. It also provides early and extensive knowledge for the authorisation process for offshore wind farms on a continual basis; a process which is likewise continuously "learning". The promising combination of rather basic research of the effects, broadly designed surveys for a better understanding of the ranges, for example of marine mammals, and the development of transfer knowledge and methods is particularly appreciative. The latter refers to research contributions designed to translate the gained basic scientific data into information suitable for operations and relevant for the decision-making process. Of course, a long-term uncertainty of the knowledge base is likely, considering the required impact prognoses and the foresight which is always involved with environmental assessments. This also holds true for the risk assessments of ship collisions, which have also been examined.

From the development of inland wind energy we know that with scale leaps from niche production to a large-scale technology, which is particularly pronounced in the offshore area, renewable energies will also have to face the question of whether the desired developments may involve a conflict of goals, for instance with the area of conservation. This question must be seriously addressed. Otherwise, concerns originated in individual cases could rapidly take on a "life of their own" and significant problems of public acceptability could develop.

Just as well, some European countries such as Denmark or Great Britain have made their contribution to environmental precaution by conducting environmental investigations of offshore wind parks at sites near the coasts. This book also provides an initial overview of the relevant European context, covering Denmark, Great Britain, the Netherlands and Sweden. Initial results of *ex-post* examinations (monitoring) at the Danish sites supplement the prior examinations at the German sites in the Exclusive Economic Zone in the North and Baltic Sea.

With the construction of a research platform, Germany is making an essential contribution to a further target-oriented collection of data in the context of accompanying research.

Last but not least, these early examinations have already led to an initial reordering of priorities in the environmental relevance of the impacts of offshore wind farms discussed. For instance, concerns about the effects of electromagnetic fields of submarine cables as well as other impacts on the benthos and the fish were put into perspective. At the same time, we do not deny the existence of gaps in the research and continue to make the necessary research profile tangible.

The project has been jointly carried out by the responsible ministry in Germany (the Federal Ministry for the Environment, Nature Conservation and Nuclear Safety), the relevant federal agencies (the Federal Nature Conservation Agency and the Federal Research Centre for Fisheries), the Project Management Organisation (the Jülich Research Centre) universities and their associated institutes (the Christian-Albrechts-University of Kiel and its Research and Technology Centre, the Ruhr University Bochum, the University Rostock, the University of Hannover, the Carl von Ossietzky University of Oldenburg, the Hamburg University of Technology, the Berlin University of Technology) as well as non-university research facilities (the German Oceanographic Museum Stralsund, the National Park Administration Schleswig-Holstein Wadden Sea, Institute of Avian Research, the Alfred Wegener Institute for Polar and Marine Research, Germanischer Lloyd WindEnergie GmbH, and the German Wind Energy Institute).

The authors wish to express their gratitude especially to the Federal Ministry for the Environment, Nature Conservation and Nuclear Safety, whose responsible program of ecological research made these exemplary investigations, approaches and assessments possible in the first place.

Berlin, March 2006   The editors

Contents

Research on Resting and Breeding Birds 117

Background...119

10   Possible Conflicts between Offshore Wind Farms and Seabirds in the German Sectors of North Sea and Baltic Sea...121

Research on Fish 145

Background...147

11   Distribution and Assemblages of Fish Species in the German Waters of North and Baltic Seas and Potential Impact of Wind Parks 149

List of Authors

Expanding Offshore Wind Energy Use in Germany

Udo Paschedag Cornelia Viertl Dr. Guido Wustlich	Federal Ministry for the Environment, Nature Conservation and Nuclear Safety (BMU) Referat KI I 3: Wasserkraft und Windenergie Dienstsitz Berlin: Alexanderplatz 6 10178 Berlin, Germany
Michael Heugel	Federal Ministry for the Environment, Nature Conservation and Nuclear Safety (BMU) Referat N I 5: Recht des Naturschutzes und der Landschaftspflege Dienstsitz Bonn: Robert-Schuman-Platz 3 53175 Bonn, Germany
Thomas Merck	Federal Agency for Nature Conservation (BfN) Field Office Vilm 18581 Putbus, Germany
Dr. Joachim Kutscher	Jülich Research Centre The Project Management Organisation Jülich (PtJ) 52425 Jülich, Germany

Research on Marine Mammals

Anita Gilles Klaus Lucke Dr. Meike Scheidat Stefan Ludwig Dr. Ursula Siebert	Research and Technology Center (FTZ) Christian-Albrechts-University of Kiel Hafentörn 1 25761 Büsum, Germany
Dr. Harald Benke Christopher G. Honnef Ursula K. Verfuß	German Oceanographic Museum Katharinenberg 14-20 18439 Stralsund, Germany
PD Dr. Guido Dehnhardt Dr. Wolf Hanke	Department of General Zoology and Neurobiology Ruhr-Universität Bochum 44780 Bochum, Germany

Prof. Dr. Dieter Adelung
Dr. Gabriele Müller
Mandy A. M. Kierspel

Leibniz-Institute of Marine Sciences
IFM-GEOMAR Kiel
Düsternbrooker Weg 20
24105 Kiel, Germany

Nikolai Liebsch
Prof. Dr. Rory P. Wilson

School of Biological Sciences
University of Wales Swansea Singleton Park
Swansea SA2 899, Wales, UK

Barbara Frank

National Park Administration
Schleswig-Holstein Wadden Sea
Schlossgarten 1
25832 Tönning, Germany

Research on Resting and Breeding Birds

Dr. Jochen Dierschke
Dr. Klaus-Michael Exo
Elvira Fredrich

Institute of Avian Research
"Vogelwarte Helgoland"
An der Vogelwarte 21
26386 Wilhelmshaven, Germany

Dr. Ommo Hüppop
Reinhold Hill

Institute of Avian Research
„Vogelwarte Helgoland"
Inselstation Helgoland,
PO Box 1220
27494 Helgoland, Germany

Research on Bird Migration

Dr. Volker Dierschke
Dr. Stefan Garthe
Bettina Mendel

Research and Technology Center (FTZ)
Christian-Albrechts-University of Kiel
Hafentörn 1
25761 Büsum, Germany

Research on Fish

Dr. Siegfried Ehrich
Dr. Matthias H. F. Kloppmann
Dr. Anne F. Sell

Federal Research Centre for Fisheries
Institute for Sea Fisheries
Palmaille 9
22767 Hamburg, Germany

Dr. Uwe Böttcher

Federal Research Centre for Fisheries
Institute for Baltic Sea Fisheries
Alter Hafen Süd 2
18069 Rostock, Germany

Research on Benthic Associations

Dr. Alexander Schröder
Dr. Covadonga Orejas
Tanja Joschko

Alfred Wegener Institute for Polar and
Marine Research (AWI)
Section Marine Animal Ecology
Columbusstraße
27568 Bremerhaven, Germany

Dr. Ralf Bochert
Dr. Michael L. Zettler
Dr. Falk Pollehne

Baltic Sea Research Institute Warnemünde
Seestraße 15
18119 Rostock, Germany

Technical Analyses

Gundula Fischer

Germanischer Lloyd WindEnergie GmbH
Steinhöft 9
20459 Hamburg, Germany

Dr.- Ing. Karl-Heinz Elmer
Wolf-Jürgen Gerasch

University of Hannover
Institute for Structural Analysis
Appelstraße 9A
30167 Hannover, Germany

Dr. Thomas Neumann
Joachim Gabriel

Deutsches Windenergie Institut (DEWI)
Ebertstr. 96
26382 Wilhelmshaven, Germany

Dr. Manfred Schultz - von Glahn
Dr. Klaus Betke
Rainer Matuschek

Institut für technische und angewandte Physik
GmbH (itap)
Marie-Curie-Straße 8
26129 Oldenburg, Germany

Florian Biehl
Prof. Dr. Eike Lehmann

Hamburg University of Technology
Institute for Ship Structural Design and
Analysis
Schwarzenbergstr. 95 C
21073 Hamburg, Germany

Planning Aspects

Julia Köller
Prof. Dr. Johann Köppel
Dr. Wolfgang Peters

Berlin University of Technology
Department of Landscape Planning and
Environmental Impact Assessment
Straße des 17. Juni 145, EB 5
10623 Berlin, Germany

International Ecological Research

Elke Bruns
Ines Steinhauer

Berlin University of Technology
Department of Landscape Planning and
Environmental Impact Assessment
Straße des 17. Juni 145, EB 5
10623 Berlin, Germany

Conclusion and Perspective

Julia Köller
Prof. Dr. Johann Köppel
Dr. Wolfgang Peters

Berlin University of Technology
Department of Landscape Planning and
Environmental Impact Assessment
Straße des 17. Juni 145, EB 5
10623 Berlin, Germany

Expanding Offshore Wind Energy Use in Germany

1   Offshore Wind Energy Use

Udo Paschedag

The German government attaches great importance to the expansion of renewable energies with a view to effective climate protection, the development of a sustainable energy supply, greater independence from energy imports and the creation of new jobs. The German government's goal, laid down in the Renewable Energy Sources Act (EEG), is to increase the share of renewable energies in the energy supply to at least 12.5 % by 2010 compared with the year 2000, and to double the share to at least 20 % by 2020. With this, Germany is making an important contribution to the EU's goal of increasing the share of renewable energies in electricity consumption from 14 % (1997) to 22 % (2010). Furthermore, the German government has also set itself a long-term goal, within the framework of its Sustainability Strategy, to cover around half of energy consumption in Germany with renewable energies by the middle of this century.

In 2004 renewable energies already accounted for 3.6 % of primary energy and 9.3 % of electricity consumption. In order to achieve the German government's further goals, the potential of various renewable forms of energy must be exploited in accordance with the best-available technology.

Now, that the potential of hydropower in Germany has already been exploited to a large extent, the greatest potential for expansion up to 2020 lies in the wind energy sector, in particular in the field of offshore wind energy. There is both advanced technological development and proven experience with the technology in this field. Other renewable energy sectors also promise comparable developments, for example biomass, solar technology and geothermal power (BMU 2004).

Every form of renewable energy has to contribute to achieving the goal of doubling the renewable share. The only way it is possible to respond to different electricity needs and the related power plant structure (base, average and peak load) is with a mix of all renewable energies. This is why it is essential to use every form of renewable energies in accordance with their level of development.

1.1   Wind Energy Use in Germany

In recent years there has been rapid development in the wind energy sector (Fig. 1). This process, in which Germany has played a key role, can be observed both nationally and internationally.

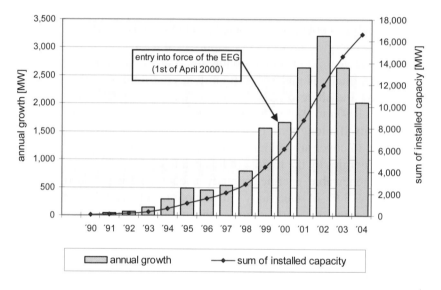

Fig. 1. Expanding the use of wind energy in Germany (Source: BWE 2005, BMU 2004)

At the end of 2004 there were 16,543 wind turbines in Germany with an output of around 16,630 MW. They were supplying 4.1 % of gross electricity consumption, corresponding to 44.8 % of electricity from renewables. In an average year, these turbines produce electricity for around 8.5 million households. In comparison with 2002 and 2003 there was a decrease in the newly installed output in 2004. This can primarily be traced to a lack of onshore sites and to as yet only small-scale repowering.

Due to the current market development and the areas designated so far as suitable sites, the potential of onshore expansion is assumed to be around 25,000 MW (Deutsche WindGuard GmbH 2005).

1.2   Potential of Offshore Wind Energy Use

In order to maintain a high level of expansion in wind energy use in Germany, the gradual development of suitable offshore sites is also necessary in addition to further expansion at suitable onshore sites and repowering. From a current perspective, the offshore areas expected to be available up to 2010 have a possible capacity of 2,000 to 3,000 MW. If economic efficiency is achieved, 20,000 to 25,000 MW of installed power is possible in the long term up to 2030 (BMU 2002) (see chapter 2).

1.3   Level of Offshore Wind Energy Use

So far only one wind turbine with an output of 4.5 MW has been installed in Germany near Emden in the territorial sea, although at a shallow depth near the dyke.

There are still considerable technical uncertainties connected to projects for offshore wind farms in the Exclusive Economic Zone (EEZ). Due to Germany's comparatively short coastline, the location of national parks in the territorial sea and socio-political concerns, the potential sites are almost exclusively those with significant water depth that are located far away from the coast. However, there is no experience in this area anywhere in the world. Project associations are carrying out pioneering work in this field.

Excluding projects already approved, there are currently 21 applications pending at the Federal Maritime and Hydrographic Agency (BSH), of which 17 are for the North Sea and four for the Baltic Sea (see Fig. 2 and Fig. 3).

The majority of these applications concern areas in the North and Baltic Seas that lie far away from the coast or the islands and have a sea depth of more than 30 metres. In addition, nine wind farms with a planned total of over 600 turbines and around 2,600 MW have been approved in the North and Baltic Seas. However, for most of these projects the licence for the cable lines from the respective Federal *Land* is still outstanding. A gradual establishment of the licensed offshore wind farms can therefore be expected from 2007 at the earliest.

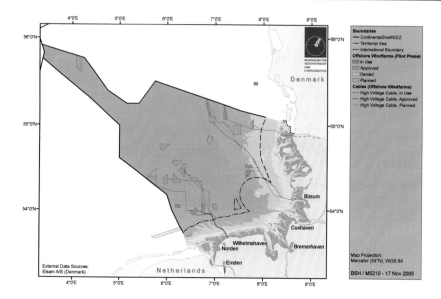

Fig. 2. North Sea: Offshore wind farms (pilot phase) (Source: BSH 2005)

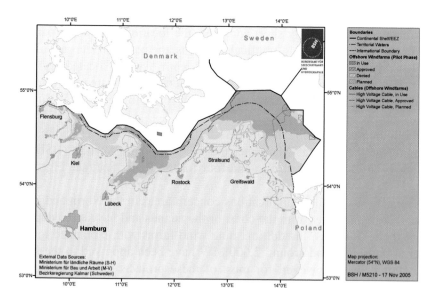

Fig. 3. Baltic Sea: Offshore wind farms (pilot phase) (Source: BSH 2005)

References

BMU (Federal Ministry for the Environment, Nature Conservation and Nuclear Safety) (ed) (2004) Erneuerbare Energien – Innovationen für die Zukunft. BMU, Berlin, 128 pp

BMU (Federal Ministry for the Environment, Nature Conservation and Nuclear Safety) (ed) (2002) Strategy of the German Government on the use of offshore wind energy in the context of the national sustainability strategy of the Federal Government. BMU, Bonn, 27 pp

BSH (Federal Maritime and Hydrographic Agency) (2005) Continental Shelf Research Information System (CONTIS) - maps

Deutsche WindGuard GmbH (2005) Kurzgutachten zum Ausbau der Windenergienutzung bis 2020

2   Strategy of the German Government

Cornelia Viertl

Within the framework of the Sustainability Strategy, the Federal Environment Ministry, as lead Ministry, elaborated a "Strategy of the German Government on the Use of Offshore Wind Energy". This strategy was published at the beginning of 2002 (BMU 2002)[1].

Its aim is to increase the share of wind energy in electricity consumption to at least 25 % over the next three decades. A 15 % share in electricity consumption in Germany is achievable with offshore wind energy alone.

Table 1. Gradual development of the use of offshore wind energy

Phases	Period	Potential capacity	Potential power yield
1. Preparational phase	2001 - 2006	-- MW	-- TWh p.a.
2. First expansion phase	2007 - 2010	2,000 - 3,000 MW	approx. 7 - 10 TWh p.a.
3. Additional expansion phases	2011 - 2030	20,000 - 25,000 MW	approx. 70 - 85 TWh p.a.

2.1   Key Elements of the Offshore Strategy

One fundamental requirement is that the expansion of offshore wind energy use is compatible with the environment and nature, and also economically viable. It is to be carried out in a step-by-step process. Prerequisites were created for the designation of protected areas and provisions for especially suitable areas for wind turbines in the Exclusive Economic Zone. Technical, environmental and nature conservation research was part of the strategy and is to accompany the expansion of offshore wind energy use for a longer period of time beyond the starting phase. In order to take due account of the precautionary principle a step-by-step expansion (first step: maximum 80 turbines) is planned. Reaching the next respective step presupposes a positive and reliable result with regard to environmental impacts.

---

[1]   see: www.erneuerbare-energien.de or www.offshore-wind.de

2.2   Successes and Focal Points for Implementation

A range of successes have already been recorded. Twelve applications for offshore wind farms have already been approved (March 2006). In doing so it was possible to enforce the step-by-step principle, according to which a maximum of 80 turbines per wind farm were licensed in the first instance. Once reliable data on the impacts on maritime navigation and the marine environment are available it will be possible to grant licences for larger wind farms.

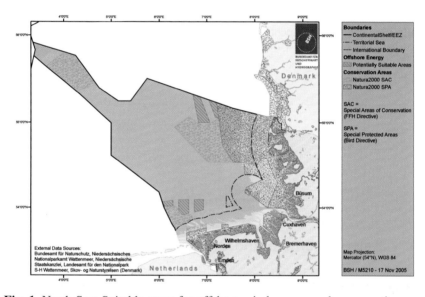

Fig. 1. North Sea: Suitable areas for offshore wind energy and conservation areas (Source: BSH 2005)

The identification and selection of German protected sites in the Exclusive Economic Zone (EEZ) for the EU Natura 2000 system has been carried out. In Mai 2004 Germany has nominated eight proposed Sites of Community Interest (pSCI) and two Special Protected Areas (SPAs) in its EEZ of the North Sea and the Baltic Sea to the EU Commission. By this step Germany covers approx. 31 % of its EEZ and by including the current nominations in the territorial seas 41 % of its total marine area by NATURA 2000 sites (see Fig. 4 and Fig. 5).

By the end of 2005 the first especially suited areas are to be designated on the basis of the Marine Facilities Ordinance (SeeAnlV). Furthermore, the legal foundations for long-term regional planning have been created by extending the Federal Regional Planning Act (ROG) to the EEZ (§ 18a), and the regional planning process has started (see Fig. 4 and Fig. 5).

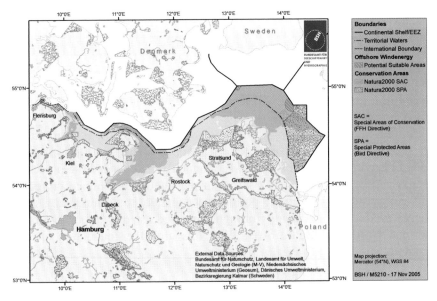

Fig. 2. Baltic Sea: Suitable areas for offshore wind energy and conservation areas (Source: BSH 2005)

A study initiated by the German Energy Agency (dena) analysed the impacts of the grid connections of offshore wind farms on the German power grid and power plant structures. This study revealed that the cost-effective integration of wind energy with a moderate grid expansion is possible, even assuming a rather high offshore wind expansion scenario (dena 2005). The structural change required for sustainable, decentralised power generation necessitates an adaptation of the German power grid. The necessary extension to the grid at extra-high voltage level identified by the study amounts to around 850 km by 2015. With a total length of the power grid in Germany of 18,000 km, this corresponds to less than five percent of the existing extra-high voltage grid. The study showed that there is no threat of critical system situations or blackouts in Germany as a result of the additional expansion of wind energy use that cannot be resolved through technical measures.

According to the changed structure of existing power plants identified by the experts, there will be no need to build additional conventional power plants (so-called shadow power plants) for the provision of balancing and reserve power. In order to conclude the first part of the grid study within the envisaged time frame, the review of further technically innovative solutions for the integration of electricity from wind energy and the optimisation of the grid expansion was postponed to a subsequent study. This study will address in particular the impacts of cable temperature monitoring, generation and feed-in management, load management and currently available storage technology such as compressed-air storage power plants. This subsequent study is expected to have repercussions on the results of the current grid study and will lead to a significant reduction in the new construction of the grid.

An Offshore Wind Energy foundation was set up by the relevant industry sectors in order to set up a test field for offshore wind turbines. The Federal Environment Ministry (BMU) launched and headed the process to establish this foundation. Its goal is, in the interest of climate protection and energy supply security, to promote sustainable, environmentally sound energy production and supply through improved use of wind energy in the German North and Baltic Seas. With this aim, the foundation aims to promote:

1. technological research, development and innovation in the field of offshore wind energy, taking account of energy transport to the consumer,
2. accompanying ecological research on the impacts of the construction, operation and decommissioning of offshore wind turbines including their cable connection on the marine environment, and
3. the exchange and transfer of knowledge on offshore wind energy between science, industry and other public and private organisations.

The foundation will acquire the rights for the licensing of an offshore wind farm that is particularly suited to the operation of a test field. Following this it will lease the sites at this wind farm to operator companies, whereby the primary goal will be the testing of multi-megawatt turbines (larger than 5 MW).

The accompanying ecological research, the research on the measuring platforms and the further development of turbine technology was secured for the long term through the Future Investment Programme (ZIP) and research programmes on renewable energies.

The first of three research platforms (FINO 1) started operation in 2003 in the North Sea approx. 45 km north of the island Borkum (Fig. 3). For further information see chapter 15.

Fig. 3. FINO 1 Research platforms in the North and Baltic Seas

The second research platform (FINO 2) is to be set up early in 2006 in the Baltic Sea near Kriegers Flak. The third platform (FINO 3/Neptun) for the northern area of the North Sea off Sylt is current at the planning stage. This publication provides an overview of the projects concerning accompanying ecological research and their results.

References

BMU (Federal Ministry for the Environment, Nature Conservation and Nuclear Safety) (ed) (2002) Strategy of the German Government on the use of offshore wind energy in the context of the national sustainability strategy of the Federal Government. BMU, Bonn, 27 pp

BSH (Federal Maritime and Hydrographic Agency) (2005) Continental Shelf Research Information System (CONTIS) - maps

dena (Deutsche Energie Agentur) (2005): Energiewirtschaftliche Planung für die Netzintegration von Windenergie in Deutschland an Land und Offshore bis zum Jahr 2020 (dena-Netzstudie).

3 Legal Framework Conditions for the Licensing of Offshore Wind Farms

Guido Wustlich, Michael Heugel

3.1 The Renewable Energy Sources Act – Support Instrument for the Expansion of Renewable Energies

In Germany, generating electricity by wind-powered plants is promoted by the Act on Granting Priority to Renewable Energy Sources (Renewable Energy Sources Act - EEG). This act is an effective and efficient instrument for increasing the use of renewable energies on the road towards a sustainable energy system. The core elements of the EEG are:

- the priority connection of installations for the generation of electricity from renewable energies and from mine gas to the general electricity supply grids;
- the priority purchase and transmission of this electricity;
- a consistent fee for this electricity paid by the grid operators, generally for a 20-year period, for commissioned installations, this payment is geared around the costs, and
- the nationwide equalisation of the electricity purchased and the corresponding fees paid.

The fee paid for the electricity depends on the energy source and the size of the installation. The rate also depends on the date of commissioning; the later an installation begins operation, the lower the tariff (degression). Since the amendment of the EEG in 2004, power from offshore wind farms will be eligible for an initial rate of 9.1 cent/kWh if the plant is commissioned by 2010 (previously 2006). Wind farms are classified as offshore if they are constructed at least three nautical miles off the shoreline. The initial rate will be paid for 12 years. This period will be extended for installations built at a greater distance from the shoreline and at greater depths. The base rate which follows the initial rate is 6.19 cent/kWh. Altogether a guaranteed price is paid for 20 years.

3.2   Licensing Offshore Wind Farms in the Territorial Sea

The territorial sea, i.e. the zone of 12 nautical miles off the German coast, belongs to the territory of the Federal Republic of Germany. The same provisions of the Federation and coastal Länder apply here as on land. Hence, the Federal Immission Control Act (BImSchG) is also applicable for the licensing of wind farms in the territorial sea.

For their construction and operation wind farms require a licence pursuant to Article 4 BImSchG (cf No. 1.6 of the Annex to the Ordinance on Installations Subject to Licensing – 4th BImSchV). In principle, it is not necessary to obtain separate further authorisations from other authorities in addition to the licence under immission control law, because the latter triggers a so-called formal concentration effect under Article 13 BImSchG and incorporates other authority decisions relating to the installation.

The licensing procedure must be carried out by the competent immission control authority of the coastal Länder as a formal procedure pursuant to Article 10 BImSchG. The procedure involves an environmental impact assessment and public participation, since offshore wind farm projects regularly concern the construction and operation of more than 20 wind turbines (cf column 1 of Annex 1 to the Act on the Assessment of Environmental Impacts in conjunction with Article 2 para. 1 No. 1 (c) of the 4th BImSchV).

The licence must be granted if it is ensured inter alia that no harmful effects on the environment may be caused and no other public law provisions oppose the construction and operation of the offshore wind farm (cf Article 6 para. 1 in conjunction with Article 5 BImSchG). Prior to the licence being granted, therefore, there must be an examination into whether the project complies with the regulations of the relevant Land building code, and whether interference with the safety and easy flow of shipping as defined in the Federal Waterways Act can be ruled out. Particular importance is attached to the concerns of nature protection and landscape management. The project must be an admissible intervention in nature and landscape. In the territorial waters of the North Sea which have been designated by Land legislation as Wadden Sea National Parks, the construction and operation of an offshore wind farm is only permissible in exceptional cases.

3.3   Licensing of Offshore Wind Farms in the Exclusive Economic Zone (EEZ)

The legal situation in the Exclusive Economic Zone, which covers the marine area beyond the territorial sea, is more complex. The EEZ does not belong to the national territory of the Federal Republic of Germany. Prevailing opinion maintains that national law which applies in the national territory is only applicable in the EEZ if the legislator has expressly declared it to be so. However, the 1982 United Nations Convention on the Law of the Sea which entered into force in 1994 grants coastal states certain utilisation privileges and regulatory powers in these marine areas, including specifically for the construction and operation of installations for the generation of energy from wind (cf Articles 56 and 60). The installations must not interfere with the use of recognised sea lanes essential to international navigation. Due notice of the construction must also be given and permanent warning systems maintained. Furthermore the coastal states can set up reasonable safety zones around the installations in order to ensure the safety of navigation and of the installation itself.

Besides the United Nations Convention on the Law of the Sea there are numerous other international agreements which contain individual regulations on the protection of marine environments or which are generally geared to these concerns. The most important of these for the Federal Republic of Germany, as a contracting party, are the two regional agreements on the protection of the marine environment of the North-East Atlantic and of the Baltic Sea of 1992 – the OSPAR Convention (in force since 1998) and the Helsinki Convention (in force since 2000). While the two conventions and the decisions and recommendations adopted on their basis contain provisions for offshore oil rigs, they do not as yet stipulate any specific requirements for offshore wind farms. However, in 2004 the OSPAR Commission published a report on this issue entitled "Problems and Benefits Associated with the Development of Offshore Wind Farms".

These conventions are directed exclusively to their respective signatory states. They only become effective within the country when they have been implemented in national law. For the construction and operation of installations in the area of the Exclusive Economic Zone this essentially takes place through the Marine Facilities Ordinance (SeeAnlV), issued on the basis of the Federal Maritime Responsibilities Act. Under Article 2 sentence 1 in conjunction with Article 1 para. 2 sentence 1 No. 1 SeeAnlV, construction, operation and essential changes to fixed or floating fixed structural or technical facilities for the generation of energy from wind require authorisation from the Federal Maritime and Hydrographic Agency (BSH).

Authorisation must be refused if the safety and easy flow of shipping is hindered or if the marine environment is endangered and this cannot be prevented or compensated for through a time limit, conditions or orders (Article 3 sentence 1 SeeAnlV). The marine environment is endangered especially if pollution of the marine environment as defined in the United Nations Convention on the Law of the Sea is to be feared or if bird migration is jeopardised. Before issuing a licence the BSH must obtain the agreement of the responsible local waterways and shipping directorate (Article 6 SeeAnlV).

The licensing procedure is based on Article 5 SeeAnlV. It is important to note that if several applications have been submitted for the same site the BSH must decide first on the application which first qualifies for a licence (so-called priority principle). This ensures that suitable sites are not blocked for years by projects which are not taken further by their applicants. Also in the procedure under the SeeAnlV an Environmental Impact Assessment must as a rule be conducted for offshore wind projects (cf Article 2a sentence 1 SeeAnlV). The licensing procedures will be simpler and quicker for offshore wind farms located in areas specially suited for wind turbines: On the basis of Article 3a SeeAnlV the competent Federal Ministries can specify areas where it can be assumed that navigational or marine environment reasons do not oppose the selection of wind farm sites. These especially suited areas laid down up to 1 December 2005 must furthermore be adopted as objectives of the spatial planning now also taking place in the Exclusive Economic Zone and laid down as priority areas (cf Article 18a para. 3 of the Federal Regional Planning Act).

Two European directives are also especially significant for the licensing eligibility of offshore wind farms: the so-called Bird Directive (Directive 79/409/EEC on the conservation of wild birds) and the Habitats Directive (Directive 92/43/EEC on the conservation and the natural habitats and of wild fauna and flora). In the opinion of the Federal Government and the EU Commission both the Habitats Directive and the Bird Directive are applicable not only within the national territory of the EU Member States but also in the marine areas directly adjacent to the territorial waters. The new Article 38 of the amended Federal Nature Conservation Act (BNatSchG) of 2002 therefore empowers the Federal Ministry for the Environment, Nature Conservation and Nuclear Safety to select and place under protection marine areas in the EEZ and the continental shelf as parts of the European "NATURA 2000" network. This provision takes into account the restricted regulatory possibilities arising for the Federal

Republic of Germany as a coastal state in view of the rights of other states in these marine regions established under international sea law. For projects on the generation of energy from water, currents or wind, it prescribes the appropriate application of Article 34 BNatSchG. Thus, prior to the licensing of an offshore wind farm, insofar as this is deemed capable, either individually or in conjunction with other projects or plans, of considerably impairing a protected marine area, the project shall be reviewed for compatibility with the respective conservation objectives. If the results of this review are negative the wind farm can only be licensed under the strict conditions of Article 34 paras. 3 to 5 BNatSchG. However, other special provisions of the BNatSchG are not applicable in the Exclusive Economic Zone, in particular the intervention and compensation regulations of Articles 18 et sqq. of the BNatSchG.

A further important step towards implementing the obligations under community law was achieved with the notification of a total of eight proposed sites under the Habitats Directive and two bird protection areas in the North and Baltic Seas. This was prepared by the Federal Nature Conservation Agency and concluded in May 2004 by the Federal Ministry for the Environment, Nature Conservation and Nuclear Safety. This protection is reinforced by the Renewable Energy Sources Act (EEG): under the new Article 10 para. 7 of the amended EEG of 2004, the payment rates under the EEG will no longer apply to electricity from wind energy installations whose construction was licensed after 1 January 2005 in an area in the Exclusive Economic Zone or the territorial sea which is already a declared or at least notified protected area under the Habitats or Bird Directives. This provision also supplements the instrument for especially suited areas under Article 3a of the SeeAnlV (see above) and directs the expansion of offshore wind energy towards areas which are unobjectionable in terms of nature protection.

In addition to the licensing of offshore wind farms, a permit must also be granted for laying the cable for the grid connection. For the Exclusive Economic Zone, this again requires a licence from the BSH, while in the territorial sea an authorisation from the river and shipping police, a permit under water law and – where applicable – a licence under dyke law and an exemption from the bans of the respective national park legislation are required. Responsibility for this – as generally in the territorial sea (see above) – lies within the authorities of the Länder. In spite of the division of responsibilities, an overall assessment should be undertaken during the licensing procedure, i.e. the environmentally relevant impacts of the offshore wind park including cable connection should be jointly evaluated.

References

OSPAR Commission (2004) Problems and Benefits Associated with the Development of Offshore Wind-Farms. 18 pp

4   Protection of the Marine Nature and Environment

Thomas Merck

Offshore wind energy utilisation will play an essential role in the renewables' share of Germany's total electricity generation. The aim of the Federal Government is for this utilisation to be environmentally and ecologically compatible (BMU 2002). To achieve this goal, special importance is attached to the protection of marine biodiversity and the environment in all phases of the development and expansion of offshore wind energy utilisation.

Offshore wind energy represents an emerging technology. The first offshore wind energy installations were erected at the beginning of the 1990s in Denmark and Sweden. In view of the increasing demands on large marine areas in the North and Baltic Seas made by sand and gravel extraction, oil and gas exploration, fishing etc, the potential negative effects arising from the planned large-scale wind energy projects must be carefully investigated. Knowledge of marine ecosystems and of the impacts of offshore wind energy installations is still less than that of land-based wind farms. Thus, in order to produce sound prognoses regarding the impacts of offshore wind farms on individual marine subjects of protection, and to conclusively assess their significance for nature, there must be a continuation of the studies which have now been underway for several years.

The following describes the main impacts which the construction and operation of offshore wind energy installations could have on some species and habitats of concern, especially if no measures are taken to prevent or reduce these impacts.

Offshore wind energy installations can disturb and displace resting and foraging seabirds, and for sensitive species this may result in permanent loss of habitat. In-flight collisions with installations (of both migrating and local birds) can lead to direct losses of individuals. The benthos in the immediate area of the installation's foundations is, of course, destroyed. The installation's influence on the hydrology and sediment conditions may also alter the benthic communities in the vicinity of the installation. Animal and plant species more rarely occurring in the German Bight, which otherwise is dominated by soft bottom communities, settle on the artificial hard substrate. Negative impacts on marine mammals arise from underwater noise especially due to construction but possibly also due to the operation of the wind farm. It is uncertain whether this possible deterrent effect would lead to a restricted use of habitat in wind farm areas. Such an effect cannot be ruled out for sensitive fish species as well. On the other hand, positive effects especially for fish and benthic fauna could also be expected, provided that fishery activities (including aquaculture) are effectively excluded in the wind farm area.

Artificial magnetic and/or electrical fields occur at cable connections which may interfere with the short- and long-range orientation of fishes and marine mammals. Moreover, the sediments surrounding the cable are heated and this may lead to cold-sensitive or thermophilic benthic species settling there. During construction, sediment plumes caused by cable laying and pile driving could result in local damage to fish roe or to the filtering apparatus of benthic organisms.

In clear weather conditions sites closer to the coast will be visible from the land and will thus alter the scenery.

Finally, the presence of wind farms does increase the risk of shipping collisions, which under certain circumstances – e.g. through oil or chemical leaks – could threaten very large areas located far from the actual wind farm.

Some results regarding the ecological impacts of wind turbines are now available from studies accompanying construction and operation, in particular from Denmark and Sweden. These provide initial findings and indications especially regarding the impacts on marine mammals, and resting and migratory birds (Zucco and Merck 2004). In addition to this publication a research project of the German government aimed at evaluating mainly international reports and studies, will produce an English summary report (Zucco et al. 2006).

Deterrent effects related to the construction of the wind farm were recorded in the case of harbour porpoises. After operation commenced, harbour porpoises were again sighted within the wind farm, albeit in smaller numbers. Among seabirds, reactions to offshore wind farms vary widely from species to species. While some (e.g. loons, auks) continued to avoid the wind farm and its surroundings after operation commenced, other seabird species (e.g. long-tailed ducks, various larid species) remained within the wind farm area, although sometimes in reduced numbers. It was furthermore observed that a large number of migratory birds avoided the wind farms and consequently, at least in good visibility, did not risk collision.

Because of variability across species and locations, caution should be applied in applying the results from one study to a location where species composition is not the same, or which may exhibit a different underwater sound profile because of different water depths or seabed characteristics. Due to the limited duration of studies to date, it is also not known whether long-term changes will occur or whether instead some species can become habituated. This underlines the importance of continued monitoring at existing wind farms at various locations.

Experimental studies of impacts conducted either at the installations or under laboratory conditions, will contribute necessary knowledge on effect mechanisms. Only this will allow the impacts of the planned broad-scale offshore wind energy utilisation to be better predicted, and evaluated for individual ecosystem components and species and habitats of conservation concern.

In order to protect the marine nature and environment, in the implementation of wind energy projects, areas of special ecological value will be excluded from development. To identify such areas, greater knowledge is required regarding the occurrence and spatial distribution of e.g. marine habitat types, benthic invertebrates, fish, birds and marine mammals. Such information must be available area-wide, obtained through corresponding broad-scale research. To this end, the German Government launched in 2001 a comprehensive research programme to evaluate the status of different relevant species groups and habitats, and to study possible negative effects arising from wind turbines. The findings from this played an important part in identifying and delimiting a total of ten marine protected areas under the EU Habitats and Birds Directives. These areas have now been notified to the EU Commission in Brussels. The EEG provision exempting offshore wind farms within marine protected areas from payments under the EEG is aimed at ensuring that these areas remain free from wind energy development.

In the development of offshore wind farms from a nature protection point of view, the aim should be to concentrate on just a few areas which are also with regard to nature conservation requirements, especially suited for offshore wind energy utilisation. The above mentioned research projects have produced findings for the evaluation of such potentially suitable offshore wind farm sites in the German Exclusive Economic Zone. The findings furthermore serve as a basis for evaluation of wind farm sites already applied for, or for analysing the results of the required Environmental Impact Assessments in the context of long-range distribution patterns. Due to the dynamic of marine systems and the variability between different years these research projects must be continued over adequate time periods.

During the construction of wind farms measures should be taken to allow the greatest possible prevention and reduction of negative impacts on the living and non-living marine environment. Accompanying ecological research to date has shown that technical mitigation measures (e.g. the selection of foundation or cable types, ramming procedures, cable depths, accident prevention) impacts such as noise emissions, turbidity plumes, or the development of electromagnetic fields can be prevented or reduced. The above-mentioned research on ecological impacts at existing wind farms and in the laboratory also aims to promote the development, and to test the effectiveness, of such mitigation measures. Additionally, seasonal restrictions should be imposed to ensure that construction of the wind energy installation does not take place during especially sensitive phases in the life cycle of affected species (especially marine mammals and birds). This requires an adequate knowledge of their biology, to be obtained through additional basic research.

In view of the still considerable gaps in knowledge, the precautionary principle must apply, whereby the development of wind energy utilisation in Germany's marine areas should be gradual, and in such a fashion as likely to produce the least ecological impact. To recognise possible negative impacts on the marine environment and nature at an early stage, and to be able to deal with them – perhaps in the form of suitable mitigation measures – parallel to the development of offshore wind energy, ecological studies especially at existing wind farms and their surroundings must be continued and extended to future German installations.

Cooperation and experience exchange with European neighbours will accelerate the acquisition of knowledge. An important step in this direction is the joint declaration on research cooperation between Denmark and Germany, under which the first bilateral research projects are already underway. Holders of a license to construct a wind farm in the Germany EEZ must, in accordance with the so-called Standards for Environmental Impact Assessments, comply with monitoring and assessment protocols. These data also represent a good source of information, and should be integrated with other researches. Nonetheless, it remains that basic research on the distribution and biology of marine species is necessary to inform the development and testing of suitable mitigation measures. Finally, new findings from national and foreign research should be incorporated into plans and licences for offshore wind installations on an ongoing process, and be applied as applicable to projects already constructed or approved in order to actually achieve the desired ecologically compatible and environmentally sound utilisation of this energy form.

References

BMU (Federal Ministry for the Environment, Nature Conservation and Nuclear Safety) (ed) (2002) Strategy of the German Government on the use of offshore wind energy in the context of the national sustainability strategy of the Federal Government. BMU, Bonn, 27 pp

Zucco C and Merck T (2004) Ökologische Effekte von Offshore Windenergieanlagen. Eine Übersicht zur aktuellen Kenntnislage (Stand: März 2004). Naturschutz und Landschaftsplanung 36 (9), pp 261-269

Zucco C, Wende W, Merck T, Köchling I, Köppel J (eds) (2006) Ecological Research on Offshore Wind Farms: International Exchange of Experience (Project No: 80446001). BfN Skripten 171

5 Ecological Research Initiated by the German Federal Government in the North and Baltic Seas

Joachim Kutscher

According to the Strategy of the German Government on the Use of Offshore Wind Energy issued in 2002, the expansion of offshore wind energy deployment shall be compatible with nature and the environment and shall be accompanied by ecological research and environmental monitoring. This was one reason for establishing an ecological research programme. Another reason was to improve the basic knowledge on marine ecosystems for the discussion process in the framework of the licensing procedure for offshore wind farms in the Exclusive Economic Zone. The approval or rejection of offshore wind farms is regulated by the Marine Facilities Ordinance (SeeAnlV). Beside adverse effects on the safety and efficiency of navigation there is only one other reason to reject an offshore project – adverse effects on the marine environment. Therefore, within the licensing procedure there is a strong focus on environmental aspects. Additionally, applicants have to carry out an Environmental Impact Assessment (EIA) according to guidelines of the Federal Maritime and Hydrographic Agency (BSH) before a decision is made by BSH (BSH 2003). After construction of a wind farm the owner has to carry out an effect monitoring. Both EIA and effect monitoring are focused on the area and surroundings of the wind farm. But there is a lack of basic knowledge of the marine ecosystems in the North and Baltic Seas on a large scale (see chapter 4). EIA and effect monitoring on the local scale of the wind farms have to be financed and carried out by the companies applying for an offshore wind farm site. For large scale environmental research a government funding programme Accompanying Ecological Research on Offshore Wind Energy Deployment (AERO) was issued as invitation to tender in July 2001. AERO was a part of the Federal Governments Future Investment Programme (PTJ 2002).

AERO has already provided and shall in future provide basic data and knowledge for the identification of marine nature protection areas in accordance with the Federal Nature Conservation Act as well as the identification of especially suitable areas for offshore wind energy deployment in the North and Baltic Seas.

Other fields of government supported research for offshore wind energy deployment are technology development and the construction and operation of research platforms.

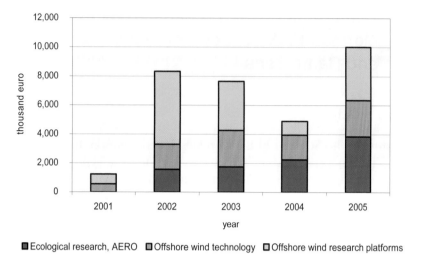

Ecological research, AERO   Offshore wind technology   Offshore wind research platforms

Fig. 1. Funds spent for offshore wind energy research (status July 2005)

Figure 1 gives an overview of how government funds for these three main fields of research have been used in the period 2001 to 2005. The total sum for this period is 32.2 million euro of which 9.5 million euro were spent on ecological research, AERO. The amount of the year 2005 in Fig. 1 represents the funds approved for projects in this year. The amounts in the year 2001 to 2004 represent the funds actually spent by the researchers.

To improve international exchange of data and experience among the neighbouring countries of the North and Baltic Seas a joint declaration was signed by the Environment Ministries of Denmark and Germany in September 2005. This declaration will strengthen bilateral know-how transfer and information exchange and facilitate joint development and implementation of projects dealing with accompanying environmental research on offshore wind energy deployment.

This publication summarises in the following chapters the results achieved so far in the framework of AERO between 2001 and 2005[1]. As Fig. 1 indicates, the funds used for ecological research (blue stripes) have been growing since AERO was announced in 2001. Main topics of the announcement of AERO in 2001 were:

---

[1] The final reports of the projects can be obtained from the project managers or from the following address: Technische Informationsbibliothek – Deutsche Forschungsberichte (TIB), Welfengarten 1 B, 30167 Hanover, Germany.

1. Impact of sound emissions and vibrations of offshore wind turbines (OWT) on marine mammals and fish, focusing on

- species and frequency related recognition ability of sounds,
- investigation of kind and intensity of the possible impact of sounds depending on frequency and intensity especially on marine mammals (communication, orientation and health),
- derivation of suitable criteria for the assessment of maximum permissible values.

2. Abundance and habitat patterns of marine mammals in the North and Baltic Seas related to the ecological relevance of potential areas for offshore wind farms or protected areas respectively

- large scale investigation of habitats of harbour porpoises, common seals and gray seals by visual, acoustic and telemetric investigation methods,
- improvement und intercalibration of detection and recording methods.

3. Bird migration and possible influence on migration paths in the North and Baltic Seas related to potential offshore wind farm areas

- large scale investigation of bird migration,
- investigation of the yearly and daily variations,
- assessment of the heights of bird migrations depending on weather conditions and other parameters.

4. Bird collisions with offshore wind turbines (OWT)

- development of detection technologies for quantitative estimation of bird collisions,
- derivation of suitable criteria for the assessment of critical values of collision frequency.

5. Time and area related dynamics of sea bird resting and reaction of resting birds to anthropogenic influence related to potential offshore wind farm areas

- investigation of the spatial distribution of resting and feeding grounds and their yearly variation,
- investigation of the influence of navigation and sensitivity of resting birds to anthropogenic influences.

6. Impact of electromagnetic fields emitted by sea cables on marine organisms

- selection of suitable physiological investigation methods for the assessment of possible influences,

- investigation of the influence of electromagnetic fields on the behaviour of marine organisms, for example the migration of fish,
- serivation of suitable criteria for the assessment of maximum permissible values of electromagnetic field strength.

7. Strategic Environmental Assessment (SEA), Environmental Impact Assessment (EIA) und Flora-Fauna Habitat Compatibility Assessment (FFH Assessment)

- development of methods and description of contents for an effective implementation of SEA, EIA and FFH Assessment for offshore wind farms.

8. Flagships of the research in the frame of AERO in the time period of 2002 to 2004 are:

- the MINOS Project coordinated by the National Park Administration Schleswig-Holstein Wadden Sea, Tönning, dealing with harbour porpoises, seals (see chapter 6, 7 and 8) and sea birds (see chapter 10);
- BeoFINO Project coordinated by the Alfred Wegener Institute for Polar and Marine Research (AWI) Bremerhaven investigating benthos and communities at the sea bottom as well as the impact of electromagnetic fields (see chapter 12, 13 and 14) and bird migration (see chapter 9);
- the investigation of sound emissions of OWT during construction (ramming of piles into the sea bottom) and during operation, development of measuring and assessment methods, prognosis of impact on marine mammals coordinated by University of Hannover (see chapter 16);
- simulation of collisions between OWT and ships and derivation of conclusions for the technical design of foundations to prevent injury of the ship body carried out by the Technical University Hamburg-Harburg (see chapter 17) and
- the requirements of Environmental Impact Assessment (EIA) and the assessment required under Article 6 of the Habitats Directive regarding the investigation and evaluation of impacts on the marine environment within the licensing procedure for offshore wind farms were worked out by the Berlin University of Technology (see chapter 18).

Most of the projects ended in 2004. After an expert workshop summarising the intermediate results and the present needs for extended knowledge in marine ecology[2] a second phase for most of the projects and

---

[2] see: www.erneuerbare-energien.de/inhalt/35811/20214/

additional projects started in 2004/2005 (cf. table in the appendix) running until 2007/2008.

5.1   Technological Research and Development

The green stripes in Fig. 1 reflect the expenditures of technological research and development projects related to offshore wind energy deployment funded by the Federal Ministry for the Environment, Nature Conservation and Nuclear Safety.

The main aspects of research were the development of large wind turbines like the 4.5 Megawatt E112 of the ENERCON company and the 5 Megawatt M5000 of the Multibrid company. Both prototypes operate at present in an onshore ore near-shore test phase. Other aspects are the integration of large volumes of wind energy into the electricity network, the improvement of wind power prognosis systems and the derivation of design parameters for offshore wind turbines and their foundations on the basis of wind and wave data measured at the offshore measuring platform FINO 1.

5.2   Platform Based Research

According to the Strategy of the German Government on the Use of Offshore Wind Energy, three offshore research platforms were to be constructed in three potentially suitable areas in close vicinity to larger offshore wind farms that are planned and applied for at the BSH. The research platforms were to be used for the following purposes:

- measuring the wind force and turbulences and their dependencies on height, wave height, the sea current and the characteristics of seabed subsoil;
- measuring the density of maritime traffic in the vicinity of the offshore research platforms;
- accompanying ecological research on issues such as migration of birds, harbour porpoise population, and benthic communities.

In the years 2002 and 2003 the first German research platform for offshore wind energy deployment FINO 1 (see chapter 15) was realised, reflected in the relatively high expenditures for platform based research (Fig. 1) in these years. FINO 1 has been in operation in the North Sea since summer 2003 with a very high availability of all measurement data.

The meteorological, hydrographical and mechanical data are collected in a database at the Federal Maritime and Hydrographic Agency (BSH) in Hamburg.

In 2005 a project started at the Schiffahrtsinstitut Warnemünde e.V. to construct a second research platform FINO 2 in the Baltic Sea about 40 km north of the island Rügen, near the German-Swedish EEZ-border in the area Kriegers Flak. In this area on both sides of the EEZ border large wind farm projects have been approved by the German and Swedish authorities. FINO 2 will be equipped with a 100 m wind measuring tower and in the field of marine ecology, with bird radar and other bird detectors, porpoise detector and equipment for the observation and monitoring of benthos and fishes. FINO 2 shall be based on a monopile construction. The deck area about 10 m above sea level will be 124 m^2. The wind measuring regime is harmonized with FINO 1 in the North Sea to obtain comparable wind data which shall be stored and managed together with the FINO 1 data at BSH. Data will be transferred onshore by radio frequency transmission so that most of the data will be available in real time. Construction of FINO 2 will be completed in 2005, depending on the weather in late autumn and winter.

A third platform is planned according to the Strategy of the German Government on the Use of Offshore Wind Energy as a research project in the North Sea, approx. 75 km west of the island of Sylt.

References

BMU (Federal Ministry for the Environment, Nature Conservation and Nuclear Safety) (ed) (2005) Erneuerbare Energien in Zahlen – Nationale und internationale Entwicklung. Stand: Juni 2005. BMU, Berlin, 47 pp

BSH (Federal Maritime and Hydrographic Agency (2003) Standards for Environmental Impact Assessment of Offshore Wind Turbines in the Marine Environment. 51 pp

BWE (Bundesverband WindEnergie), Osnabrück 2005

PtJ (Projektträger Jülich (2002) Ökologische Begleitforschung zur Offshore-Windenergienutzung. Fachtagung des Bundesministeriums für Umwelt, Naturschutz und Reaktorsicherheit und des Projektträgers Jülich, Bremerhaven 28. und 29. Mai 2002. Tagungsband. 67 pp

Research on
Marine Mammals

Background

Three marine mammal species are native to the German North and Baltic Seas: the harbour porpoise (*Phocoena phocoena*), which is native to German waters and is the most common whale species here, and two seal species, the harbour seal (*Phoca vitulina*) and the grey seal (*Halichoerus grypus*). Moreover, a number of other species of whale and seal are also encountered, although their appearance is rather rare in our waters.

Marine mammals are at the end of the food-chain in the ecosystem (top predators). Due to their size, they have no natural enemies in the North and Baltic Seas. However, the marine mammals living here are endangered by a multitude of anthropogenic effects. In addition to pollution, nutrient entry and fishing (by-catch), shipping and underwater noise represent an increasing threat to marine mammals.

The operation of offshore wind energy plants always involves acoustic emissions into the body of water, both during the construction phase and during the ensuing operational phase of the plants. Particularly during the construction of the foundations, this sound immission can be very intensive, and have lethal effects, or cause permanent damage to hearing. While the acoustic emissions only occur occasionally and temporarily during the construction phase, the noise is permanent during the regular operation of the plants. Due to the large number of individual plants, an extensive and permanent noise burden in the sea due to noise emission may arise. At present, the question as to whether these disturbing noises impair or permanently damage marine mammals is still largely unresolved. It is, however, undisputed that marine mammals are extraordinarily dependent on their hearing systems, e.g. for the intra-specific communication, for the search for food, and for orientation, and are thus particularly sensitive to noise emissions.

In the context of the planning and operation of wind parks, scientists are currently discussing the following correlations of effects, which may constitute impairments relevant for the construction and operation of offshore wind parks:

- temporary habitat loss and dislocation of the animals by construction and maintenance activities;
- permanent habitat loss due to operational noise and other activities;
- physiological damages, up to direct/indirect loss of individuals, due to construction-related noise emissions (e.g. hearing damage from ramming or drilling noise);

- disturbance of intra-specific communication (masking by noise), and a resulting reduction of the reproduction rate;
- barrier effects to migrating animals due to operation-caused noise or electromagnetic fields.

In order to assess the results by the construction and operation of off-shore wind energy plants with regard to the planning and approval of wind parks, not only data on the size and distribution of the species stock in German waters, but also audiometric data on the marine mammals themselves are needed. There is a considerable requirement for research regarding the establishment of limit values for sound emissions which could cause death or lasting damage to their hearing, and thus represent an unacceptable endangerment of the marine environment, so that the approval of a wind park would have to be rejected. The research projects on marine mammals presented below have been designed to help close these knowledge gaps.

6 Harbour Porpoises (*Phocoena phocoena*): Investigation of Density, Distribution Patterns, Habitat Use and Acoustics in the German North and Baltic Seas

Ursula Siebert, Harald Benke, Guido Dehnhardt, Anita Gilles,
Wolf Hanke, Christopher G Honnef, Klaus Lucke, Stefan Ludwig,
Meike Scheidat, Ursula K Verfuß

6.1   Introduction

The harbour, or common, porpoise (*Phocoena phocoena*) is the smallest cetacean inhabiting temperate to cold waters throughout the northern hemisphere. Due to its occurrence mainly but not exclusively in coastal or shelf waters, the porpoise is threatened by a variety of anthropogenic influences (Hutchinson et al. 1995, Kaschner 2001, Scheidat and Siebert 2003), including by-catch in fishery (Kock and Benke 1996, IWC 1997, Vinther 1999, Lockyer and Kinze 2000) and habitat degradation due to e.g. chemical pollution (Jepson et al. 1999, Siebert et al. 1999). The harbour porpoise is the only cetacean species regularly found in both the German North and Baltic Seas (Reijnders 1992, Benke and Siebert 1994, Schulze 1996, Benke et al. 1998, Hammond et al. 2002, Siebert et al. accepted).

Until recently, very little data existed on the distribution of and habitat use by harbour porpoises in German waters. Most information on distribution and population numbers in the German North and Baltic Seas was based on results of the SCANS (Small Cetacean Abundance in the North Sea and Adjacent Waters) survey of July 1994 (Hammond et al. 1995, Hammond et al. 2002). Unfortunately, the SCANS investigation did not cover some areas of the German Exclusive Economic Zone (EEZ), such as the region east of the island of Rügen close to the Polish border in the Baltic Sea, and some parts off the East Friesian Islands between the estuary of the river Elbe and the Dutch border in the North Sea.

Due to this gap in knowledge, it was necessary to investigate German waters in respect of distribution, density as well as habitat use of harbour porpoises in order to assess further anthropogenic influences and their cumulative effects, e.g. the planned construction of offshore farms. Furthermore, since noise pollution is considered a particularly important threat in this respect, more information on the animals' hearing capabilities is required to facilitate the establishment of a solid baseline for appropriate mitigation measures.

This paper summarises four parts of the MINOS[1] project which dealt with harbour porpoises:

1. Investigation of **density and distribution patterns** of harbour porpoises (*Phocoena phocoena*) in the German North and Baltic Seas.
2. Investigations on the **habitat use** of harbour porpoises in the North and Baltic Seas using autonomous echo-location click detectors (T-PODs).
3. **Intercalibration** of different methods for observing and counting harbour porpoises (*Phocoena phocoena*) in the whale sanctuary in the Nationalpark of the Schleswig-Holstein Wadden Sea.
4. Study on the **impact of offshore windmill**-related sound emissions on the auditory system of marine mammals in the German North and Baltic Seas.

6.1.1   Density, Distribution Patterns and Habitat Use

Aerial and ship-based line transect surveys were used to reveal distribution patterns and to estimate abundance of harbour porpoises in German waters. In waters of very low harbour porpoise densities, such as the western Baltic and the Baltic proper, survey results will suffer from major confidence intervals. Therefore, as an alternative method to estimate distribution and relative abundance, passive acoustic monitoring devices, so-called T-PODs (Porpoise Detectors), were deployed permanently at measuring points throughout the German Baltic Sea, from Fehmarn to the Pomeranian Bight.

The harbour porpoise, like other odontocete species, emits short pulsed high frequency click sounds for echo-location (Au 1993). As an active sensory system, echo-location in porpoises is used for both orientation and foraging (Verfuß and Schnitzler 2002, Verfuß et al. 2005). Harbour porpoise echolocation clicks are very distinct and different from most dolphin echo-location clicks (Au 1993). Their main energy is focused on a small frequency bandwidth at around 130 kHz (Goodson et al. 1995, Kamminga et al. 1999). This easily distinguishable click structure provides a good opportunity to set up an automatic system that specifically monitors this species. Taking advantage of the highly sophisticated sonar of

---

[1]   MINOS – Marine Warmblüter in Nord- und Ostsee (Marine warm-blooded animals in the North and Baltic Seas: Foundations for assessment of offshore wind farms).

porpoises, T-PODs are one approach in the development of such passive acoustic monitoring devices.

6.1.2   Intercalibration

Visual and acoustic surveys must be intercalibrated to yield a maximum of information on the occurrence and habitat use of harbour porpoises. Therefore, one of the goals of the intercalibration study was to investigate the comparability of the various monitoring methods (aerial and ship-based surveys in addition to T-PODs), by using them at the same time in the same area. Since the T-POD method is relatively new, its practical use as a routine tool for monitoring porpoises, including the traditional stationary T-POD method and tests towing T-PODs behind ships, was also investigated.

6.1.3   Impacts of Offshore Windmills

In order to assess the impact of offshore wind turbines on harbour porpoises, the increase in ship movement during construction and the associated noise in the area need to be considered. The insertion of foundations into the seabed is of special concern, as the applied technique will be accompanied by repeat emissions (>1,000 impulses per unit) of intense sound signals (225+ dB re 1 µPa at 1 m) (Ødegaard and Danneskiold-Samsøe A/S 2000). During the operational phase the windmills will continuously emit low-frequency noise into the water.

Therefore, another objective of the MINOS project was to enhance basic knowledge of the auditory sensitivity of harbour porpoises and harbour seals, to define the effect these emissions may have on the animals, and, as far as possible to quantify these potential effects. To further monitor the auditory status of animals in captivity and in the wild, a method was applied which can be conveniently used in both conditions. The electro-physiological audiometric technique of measuring the Auditory Brainstem Response (ABR) was chosen as the most suitable method for this study. It is a non-invasive technique, and has already been widely adopted with human patients. The ABR method enables measurements of auditory sensitivity even in situations in which the subject is unwilling or unable to participate in normal behavioural testing, as it requires only minimal (active) co-operation. This technique is based on the presentation of an acoustic stimulus while the evoked neural responses (Auditory Evoked Potentials, AEPs – a more general term for ABRs) are recorded by means of electrodes attached to the skin of the subjects' scalp. ABRs reflect the neuronal activity within the auditory pathway and are thus useful for measuring the functioning of the auditory system. The acoustic part of the MINOS project was designed to provide baseline data on the acoustic sensitivity and tolerance of the auditory system of harbour porpoises and harbour seals held in captivity, as well as from wild-caught harbour seals. The expected results will have ecological, behavioural and evolutionary relevance and may be analysed under a variety of aspects.

6.2   Methods

6.2.1   Density and Distribution Patterns

Aerial surveys to estimate abundance and investigate distribution patterns of harbour porpoises in the German North and Baltic Seas were conducted from May 2002 to October 2003 as part of the MINOS project. Surveys followed standard line-transect methodology for aerial surveys (Hiby and Hammond 1989, Buckland et al. 2001). The study area included the German EEZ in the North and Baltic Seas, as well as the 12 nm zone off the coastline. In the Baltic Sea, the study area extended to Danish waters for methodological reasons. Thus, the northern boundary of the area was determined by the inner Danish islands. Four survey blocks were designed in the North Sea and three were surveyed in the Baltic Sea (Fig. 1).

6.2.2   Habitat Use

T-PODs are self-contained data loggers for cetacean echo-location clicks[2], consisting of a hydrophone, filter and memory (Fig. 2). They register, in a 10 μsec resolution, the presence and length of high frequency click sounds which match specific criteria. Data logging is continuous for 24 hours a day over a period of eight to ten weeks. After this period, the data are downloaded to a computer and batteries must be replaced.

---

[2]   for details see: www.chelonia.demon.co.uk/PODhome.html

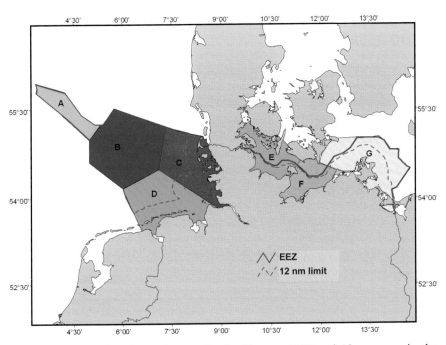

Fig. 1. Map showing the study area in the German EEZ and 12 nm zone in the North and Baltic Seas. Area A= Entenschnabel; B= Offshore; C= Nordfriesland; D= Ostfriesland, E= Kieler Bucht, F= Mecklenburger Bucht, G= Rügen. Map projection: Mercator

Fig. 2. A T-POD moored under water

T-POD Application

Twenty-one measuring positions were selected to monitor the German Baltic Sea from Fehmarn to the Pomeranian Bight (Fig. 3) from August 2002 until December 2003. At each measuring position, one T-POD was deployed on a mooring, fixed five to seven meters below the water surface. T-POD versions 2 and 3 were used. The mooring consisted of a 30-kg anchor connected to several surface buoys by a rope (Fig. 2). The listening criteria of the T-PODs are described in detail in Verfuß et al. 2004a and 2004b.

The T-PODs were calibrated before deployment to set the minimum receiving level of each T-POD. This is the level at which the device will start to register porpoise clicks. The minimum receiving level of the deployed T-PODs was in the range of 117 dB re 1 $V_{(pp)}/\mu Pa$ up to 144 dB re 1 $V_{(pp)}/\mu Pa$.

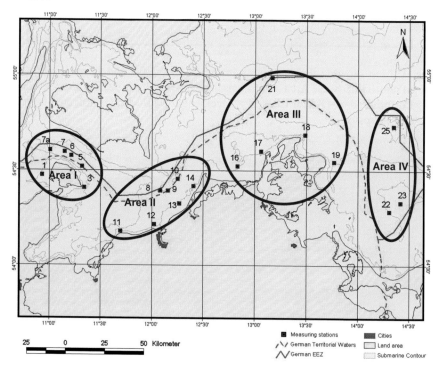

Fig. 3. Locations of all utilised T-POD-measuring stations in the Baltic Sea. The area of investigation was divided into four sub-areas (black circles): Area I: stations 1, 3, 5 - 7a; Area II: stations 8 - 14; Area III: stations 16 - 19, 21; Area IV: stations 22, 23, 25

T-POD Data Analysis

The click sounds registered from the T-PODs were scanned for trains of clicks with a specific signal pattern by means of a train detection algorithm (V2.2) included in the T-POD software. Click trains classified by the algorithm as "high probability cetacean click trains" up to "very doubtful trains" could originate from harbour porpoises, as well as from boat (e.g. sonar, propeller noise) or background noise. Therefore, these click trains were manually reviewed for harbour porpoise echo-location click trains as described in Verfuß et al. (2004a and 2004b). Only click trains classified by the algorithm and attributed manually to porpoise origin were included in the data set. Those manually attributed to other sources were rejected.

For further analysis, porpoise-positive days – defined as days with at least one classified porpoise click train – were determined from all data recordings. The percentage of porpoise-positive days among the total of monitored days per month was calculated for each position. Months with less than five monitoring days were ignored.

The monitored area of the German Baltic Sea was divided into four sections with the following T-POD positions:

- **Area I:** positions 1 to 7: western part of the German Baltic Sea, area around Fehmarn island;
- **Area II:** positions 8 to 14: western part of the German Baltic Sea, Kadet Channel and adjacent coastal area;
- **Area III:** positions 16 to 21: eastern part of the German Baltic Sea, area north of Darss and around Rügen island, incl. EEZ;
- **Area IV:** positions 22 to 25: eastern part of the German Baltic Sea, Pomeranian Bight.

The mean of the percentages of porpoise-positive days per month from the included positions was calculated for each of the four areas.

6.2.3 Intercalibration

Ship surveys, conducted according to standard line transect methodology described in detail in Buckland et al. (2001), were carried out mainly in the whale sanctuary west of the island of Sylt (Fig. 4). The methods used for the aerial surveys are described above.

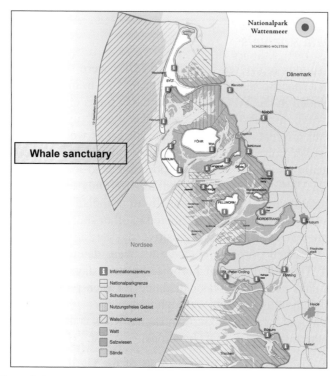

Fig. 4. The whale sanctuary of Sylt and Amrum (crossed blue) (Map modified from the national park county)

T-PODs were used in two different ways:

1. Stationary T-PODs were deployed close to the shore of the town Westerland located in the centre of the island of Sylt. They were anchored with a trapeze construction specifically developed for this project. This trapeze insured that the T-POD was vertically suspended in all water/weather conditions (Fig. 5).

2. To test the hauling of the T-POD behind a ship, it was necessary to construct a set of wings to prevent the T-POD from rotating in the water (see Fig. 6). A multi-channel data logger (Multisensor-VHF tag©, Habit Research Ltd. 1999) was mounted on the T-POD during the initial runs to test the wing construction. This device registered pressure, angle, temperature and light every two seconds and was used to evaluate the stability of the T-POD's position during towing.

Data pertaining to the following parameters were gathered from the T-PODs using the Chelonia TPOD software (August 2003 upgrade):

- porpoise-positive days (days when at least one click train from harbour porpoises was recorded),
- porpoise-positive hours (hours when at least one click train from harbour porpoises was recorded).

The exact criteria and definitions for the acoustic data are described above (see *T-POD data analysis*).

The software applications Statistica (Statsoft, Inc. 2000) and STATeasy (2001) were used for all statistical analyses.

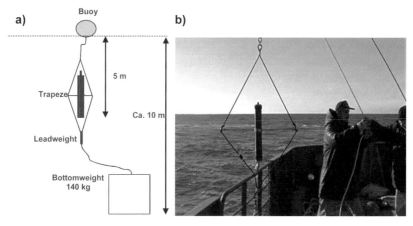

Fig. 5. Drawing on T-POD with anchoring system (a) and picture of T-POD suspended in the trapeze before deployment (b)

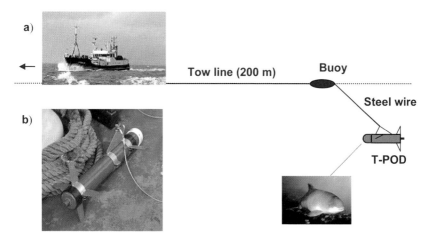

Fig. 6. Drawing of towed T-POD system behind the research vessel Littorina (a) and T-POD with wings mounted (b)

6.2.4   Impact of Offshore Windmills

An eight-year-old male harbour porpoise, held at the Fjord & Belt – Go Underwater Park in Kerteminde, Denmark, was chosen for the auditory baseline study. As an evoked potential technique for measuring hearing sensitivity was used, the animal was trained to accept two surface electrodes mounted into suction cups to be placed behind its blowhole and near the dorsal fin. With the electrodes attached in this way, it then dived to an underwater station where the acoustic stimuli were presented. By using amplitude-modulated sounds at frequencies at and above 2 kHz and sinusoidal signals at and below 2 kHz, the absolute hearing threshold of this harbour porpoise was measured at frequencies from 0.3 to 22.4 kHz in half-octave steps. The background noise level was monitored at the position where the animal received the signals and incorporated in the analysis of the results. If the sound level rose well above the average background noise level (due to e.g. ship activity in the harbour area), the tests were stopped.

Using a similar technique as described for the porpoise, a nine-year-old male harbour seal held at the Cologne Zoo in Germany was tested for its aerial acoustic sensitivity. Three electrodes were attached to the animal's skin with medical glue: the active electrode near the auditory meatus, the reference electrode on the vertex of the animal's skull, and a ground electrode on its back. All stimuli administered were short sinusoids at frequencies varying in half-octave stages between 0.125 kHz and 16 kHz. Except for 250 Hz, all frequencies could be tested repeatedly to increase the declarative strength of the results. In addition to the experiments under controlled conditions, AEP (Auditory Evoked Potential) measurements with the same acoustic stimuli as used on the captive animal were conducted on a wild-caught male harbour seal.

The ABRs following each signal presentation were recorded from the surface of the animals' skin using suction cup electrodes. These electrode responses served as input to a low noise amplifier. The amplified analogue signals were then passed through an anti-aliasing filter and led to an A/D converter. The digitised response was digitally filtered, tested for the presence of unwanted signal artefacts and averaged. Threshold values were analysed differently for the two types of signals: For the amplitude-modulated signals, an FFT analysis was first carried out on the responses before a regression analysis could be applied to the resulting values, whereas a regression analysis could be applied directly to the peak-to-peak amplitude of the ABRs. Generation of the acoustic stimuli as well as digitisation and recording of the evoked responses were performed at a TDT workstation (Tucker-Davis Technologies, USA – System 3). Stimuli were either presented via headphone for airborne sounds or via various hydrophones for underwater stimulus presentation. During all experiments, silver disk electrodes were used. The animals held in captivity had to be trained to participate in the experiments as the study design used measurements under controlled conditions.

6.3   Results

6.3.1   Density and Distribution Patterns

Summer distribution (May to August) in the German North Sea for 2002 and 2003 showed an uneven distribution of porpoises throughout the German Bight (Fig. 7). Highest densities were observed in area C around the so-called 'Amrum Aussengrund' which includes the area off the island of Sylt close to the Danish border. A north-south density gradient was observed along the coastline. The lowest densities were found in area D, off the East Friesian Islands. The aerial surveys revealed that harbour porpoises were present in the farthest reaches of the German EEZ (Doggerbank), although in slightly lower densities than at the 'Amrum Aussengrund'. It is important to note that these are the results of the summer survey (May to August) only. In addition, the overall distribution pattern still includes some areas that have not been well surveyed, due to unfavourable weather conditions, e.g. the southern part of area B (Offshore).

In contrast to the situation in the North Sea, distribution patterns in the German Baltic Sea varied profoundly from year to year (Fig. 8). In 2002, highest densities were found in the eastern part of the Baltic Sea (Pomeranian Bight), while in 2003 most animals were recorded in the western part, in area E, called 'Kieler Bucht'. Area E was not surveyed in the summer of 2002. All sightings east of the island of Rügen occurred in May and July 2002, despite considerable survey efforts during other months.

This last finding is especially noteworthy, as the population east of the underwater Darss-Limhamn ridge is considered to belong to a different population than the rest of the Baltic/Belt Sea animals (Tiedemann et al. 1996, Börjesson and Berggren 1997, Huggenberger et al. 2002). Most recent abundance estimates for this subpopulation arrived at 599 porpoises (CV = 0.57) (Berggren 1995). Joint activities of ASCOBANS (Agreement on the Conservation of Small Cetaceans of the Baltic and North Seas) and the IWC (International Whaling Commission) have stressed the precarious situation in which the stock seems to be (e.g. implementation of the 'Recovery Plan for Baltic Harbour Porpoises: Jastarnia plan').

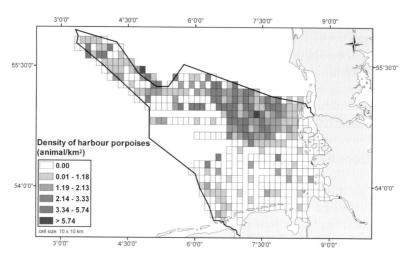

Fig. 7. Map showing the density distribution of harbour porpoises in the German EEZ of the North Sea. Density is shown as animals per km² per cell (10x10 km). All flights conducted under good or moderate conditions from May to August 2002 and from May to August 2003 are shown. Map projection: Mercator

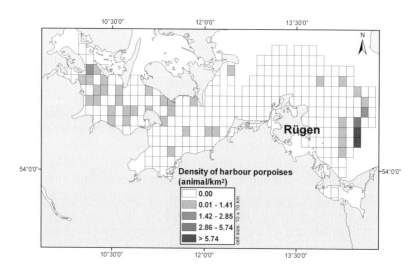

Fig. 8. Map showing the density distribution of harbour porpoises in the German EEZ of the Baltic Sea. Density is shown as animals per km² per cell (10x10 km). All flights conducted under good or moderate conditions from May to August 2002 and from May to August 2003 are shown. Map projection: Mercator

The abundance of harbour porpoises in the areas surveyed in the North and Baltic Seas was calculated as a geometric mean for the summer months, May to August (Table 1). These abundance estimates did not show major variations between the study years (34,381 harbour porpoises in 2002 and 39,115 harbour porpoises in 2003) in the German North Sea. However, in the Baltic Sea a major divergence in the abundance of harbour porpoises from year to year was calculated (Table 1). In 2002 the geometric mean summer abundance was estimated to be 4,564 animals, whereas in 2003 geometric mean summer abundance estimates arrived at a figure of 1,638 harbour porpoises. The large coefficient of variation mirrors the large difference between the years.

Table 1. Density of porpoises from May to August in the German North and Baltic Seas in the years 2002 and 2003

Area	Size (km²)	Effort (km²) 2002	No. porp.	Density (no./km²) 2002	Effort (km²) 2003	No. porp.	Density (no./km²) 2003	Abundance per area 2002	Abundance per area 2003	Mean (GM) abundance 2002 and 2003	CV
A	3,903	3.90	4	1.03	110.33	90	0.82	4,003	3,184	3,570	0.16
B	11,650	56.36	33	0.59	58.06	42	0.72	6,821	8,427	7,582	0.15
C	13,668	231.31	353	1.53	379.35	703	1.85	20,859	25,329	22,986	0.14
D	11,824	179.69	41	0.23	97.88	18	0.18	2,698	2,174	2,422	0.15
sum North Sea	41,045	471.26	431		645.62	853		34,381	39,115		
E	4,696	-	-	-	110.46	29	0.26	-	1,233	-	-
F	7,248	151.70	20	0.13	214.44	12	0.06	956	406	623	0.62
G	10,990	179.68	59	0.33	143.58	0	0	3,609	0	59	43.19
sum Baltic Sea	22,934	331.38	79		468.48	41		4,564	1,638		

The coefficient of variation (CV) was calculated for each area using the values of 2002 and 2003 as samples. The mean shown is the geometric mean (GM) based on log-transformed data. Area E (Kieler Bucht) was not surveyed in summer 2002.

6.3.2 Habitat Use

Table 2 gives an overview of the number of monitored days per month and the corresponding percentage of porpoise-positive days per month for each T-POD position. None of the positions were monitored for the entire time for logistical reasons, and in some cases because of loss of moorings. The total number of observation days is indicated.

The results show a geographical as well as a seasonal variation in the percentage of porpoise-positive days of the total number of days on which data were obtained (Fig. 9).

Table 2. Number of days monitored (obs days) and percentage of porpoise-positive days per month (% pp days) for the monitoring period from August 2002 to December 2003 for all utilised T-POD-measuring stations (1 to 25) in the German Baltic Sea, as well as average percentage of porpoise-positive days per month for area I (stations 1 - 7) and area II (stations 8 - 14). The number of stations included in the average calculation is given as "n"

	month	data	1	3	5	6	7a	7	average	8	9	10	11	12	13	14	average
						Area I							Area II				
year 2002	8	total							n = 0	31	31	31		12		31	n = 5
		%pp								64.5	64.5	77.4		16.7		83.9	67.6%
	9	total					19		n = 1	30	30	30	18	30	18	15	n = 7
		%pp					100		100%	80.0	73.3	66.7	77.8	16.7	83.3	66.7	64.3%
	10	total	9	9	23		17		n = 4	31	22	31	19	5	16	16	n = 7
		%pp	100	100	95.7		100		98.3%	58.1	45.5	64.5	52.6	20.0	62.5	62.5	56.4%
	11	total	30	30	18	18			n = 4	30			8				n = 2
		%pp	96.7	90	100	94.4			94.8%	73.3			12.5				60.5%
	12	total	31	17	18	18			n = 4	31			31				n = 2
		%pp	41.9	52.9	88.9	94.4			65.5%	19.4			0.0				9.7%
	2002 total	total	70	56	59	36		59	n = 280	153	83	96	76	47	34	66	n = 7
		2002 %pp	72.9	80.4	94.9	94.4		88.1	85.0%	58.8	62.7	69.8	32.9	17.0	73.5	71.2	56.6%
year 2003	1	total	13						n = 1	31			11				n = 2
		%pp	38.5						38.5%	3.2			9.1				4.8%
	2	total	26						n = 1	28							n = 1
		%pp	23.1						23.1%	3.6							3.6%
	3	total	31						n = 31	12		14				14	n = 3
		%pp	51.6						51.6%	16.7		7.1				0.0	7.5%
	4	total	30		16	16	16		n = 4			30				30	n = 2
		%pp	100		100	100	100		100%			6.7				0.0	3.3%
	5	total	31		31	31	31		n = 4			31				16	n = 2
		%pp	100		100	100	100		100%			12.9				31.3	19.1%
	6	total	20		30	30	30		n = 4			30					n = 1
		%pp	100		100	100	96.7		99.1%			43.3					43.3%
	7	total	14	31					n = 2	11		31	11			17	n = 4
		%pp	100	96.8					97.8%	54.5		61.3	36.4			70.6	58.6%
	8	total	31	31	17				n = 3	31		31	15			23	n = 4
		%pp	100	90.3	94.1				94.9%	48.4		77.4	53.3			82.6	66.0%
	9	total	30	30	30				n = 3	29	12	28				13	n = 4
		%pp	100	100	96.7				98.9	72.4	16.7	78.6				53.8	63.4%
	10	total	7	31	30				n = 3	31	31	31	9			31	n = 5
		%pp	100	100	100				100%	80.6	16.1	74.2	77.8			67.7	60.9%
	11	total		31	30				n = 2	30	29	30				30	n = 4
		%pp		100	100				100%	30.0	31.0	90.0				50.0	50.4%
	12	total		31	31				n = 2	31	31	31				31	n = 4
		%pp		96.8	100				98.4%	19.4	12.9	41.9				3.2	19.4%
	2003 Total	total	233		261	217	77		n = 4	234	103	287	46			205	n = 5
		2003 % pp	81.5		98.1	99.1	98.7		93.5%	36.8	19.4	51.6	43.5			39.0	40.5%

T-POD-station

Table 2 (cont.) Number of days monitored (obs days) and percentage of porpoise-positive days per month (% pp days) for the monitoring period from August 2002 to December 2003 for all utilised T-POD-measuring stations (1 to 25) in the German Baltic Sea, as well as average percentage of porpoise-positive days per month for area III (station 16 - 21) and area IV (station 22 - 25). The number of stations included in the average calculation is given as "n"

| | | | | | | | T-POD-station | | | | |
| | | | Area III | | | | | | Area IV | | |
month	data	16	17	18	19	21	average	22	23	25	average
8	total						n = 0				n = 0
	%pp										
9	total		13		5		n = 2				n = 0
	%pp		61.5		0.0		44.4%				
10	total		30	10	31		n = 3				n = 0
	%pp		33.3	30.0	0.0		18.3%				
11	total			30	7		n = 2		15	16	n = 2
	%pp			0.0	0,0		0.0%		0.0	0.0	0.0%
12	total			31		19	n = 2		26	31	n = 2
	%pp			22.6		5.3	16.0%		0.0	3.2	1.8%
2002 total			43	71	43	19	n = 4		41	47	n = 2
2002 %pp			41.9	14.1	0.0	5.3	16.5%		0.0	2.1	1.1%
1	total			31		31	n = 2		31	31	n = 2
	%pp			9.7		22.6	16.1%		0.0	6.5	3.2%
2	total			28		28	n = 2		28	28	n = 2
	%pp			3.6		14.3	8.9%		0.0	0.0	0.0%
3	total			18		10	n = 2		31	31	n = 2
	%pp			11.1		0.0	7.1%		0.0	0.0	0.0%
4	total		5	30		30	n = 3		30	30	n = 2
	%pp		0.0	0.0		0.0	0.0%		0.0	0.0	0.0%
5	total	6	31	31		31	n = 4		31	31	n = 2
	%pp	0.0	0.0	3.2		0.0	1.0%		0.0	0.0	0.0%
6	total	30	30	29		30	n = 4		30	11	n = 2
	%pp	23.3	13.3	13.8		0.0	12.6%		0.0	0.0	0.0%
7	total	31	20	31		31	n = 4	16	14		n = 2
	%pp	29.0	10.0	6.5		3.2	12.4%	0.0	0.0		0.0%
8	total	27	28	31		31	n = 4	31			n = 1
	%pp	40.7	39.3	19.4		9.7	26.5%	3.2			3.2%
9	total					30	n = 1	30			n = 1
	%pp					3.3	3.3%	0.0			0.0%
10	total			8		23	n = 2	12	18		n = 2
	%pp			0.0		0.0	0.0%	0.0	0.0		0.0%
11	total	30		30		17	n = 3		30	30	n = 2
	%pp	10.0		0.0		5.9	5.2%		0.0	0.0	0.0%
12	total	31					n = 1		31	31	n = 2
	%pp	6.5					6.5%		0.0	0.0	0.0%
2003 Total		158	11	268		292	n = 4	89	274	227	n = 3
2003 % pp		20.3	14.4	7.1		5.8	10.2%	1.1	0.0	0.9	0.5%

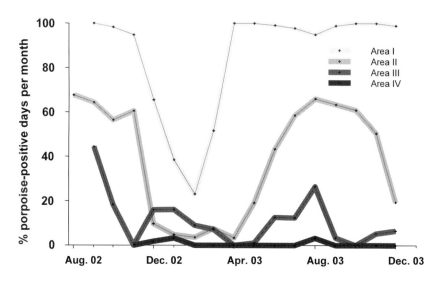

Fig. 9. Mean percentage of porpoise-positive days per month for area I to area IV over a one-year period (August 2002 to December 2003). Measurement stations included: area I (yellow): stations 1, 3, 5 - 7a; area II (orange): stations 8 - 14; area III (red): stations 16 - 19, 21; area IV (dark red): stations 22, 23, 25

In area I (area around Fehmarn island), the average percentage of porpoise-positive days per month was around 100 % from September to November 2002 and April to December 2003. It dropped to 66 % in December 2002, to 39 % in January 2003, and to a minimum of 23 % in February 2003. In March 2003 the average percentage of porpoise-positive days per month rose to 52 %.

In area II (Kadet Channel and adjacent coastal areas), the average percentage of porpoise-positive days per month was above 70 % in August and September 2002; it declined to below 10 % for December 2002 through April 2003, and rose again above 60 % in August 2003, dropping again below 60 %, to 50 % and 19 % in November and December 2003, respectively.

In area III (area north of Darss and around Rügen island, incl. EEZ), the average percentage of porpoise-positive days per month started with 44 % in September 2002, dropped and stayed below 20 % from November 2002 to July 2003, with the lowest values in November 2002 and April/May 2003, and increased during the winter months of 2002/2003. In August 2003, the average percentage of porpoise-positive days peaked at 27 % and dropped again to values below 10 % for the remaining months of the year.

In area IV (Pomeranian Bight), the average percentage of porpoise-positive days per month was near 0 %, with one or two porpoise-positive days in December 2002, as well as in January and August 2003 (resulting in up to 3 %).

6.3.3 Intercalibration

Aerial and Ship Surveys

On two occasions, it was possible to conduct ship and aerial surveys at the same time. A third ship survey was cut short due to inclement weather and did not cover the entire survey area. Harbour porpoise observations from these surveys are shown in Table 3.

An estimate of $g(0)$ based on the ship surveys was not possible because the low number of sightings rendered the estimated number of porpoises in the sanctuary unreliable. A direct comparison between ship and plane data was thus impossible. The sightings/km and animals/km were low, both for ship the and for the aerial surveys, which is consistent with previous findings (Caretta et al. 2001). More surveys are needed to confirm these findings. More sightings of animals during each survey are needed to estimate $g(0)$ and allow for direct statistical comparison of the numbers of animals in the whale sanctuary.

Table 3. Overview of sightings from concurrent ship and airplane surveys in the whale sanctuary ($1.74 \ km^2$)

Date	No. of sightings	No. of animals	No. of calves	Sightings/ km	Animals/ km	$\varnothing$ group size	Population without $g(0)$	Estimate with $g(0)$
2002								
Airplane (16/8)	23	28	1	0.10	0.13	1.2	168	293
Ship (14-16/8)	8	13	0	0.03	0.04	1.6	-	-
2003								
Airplane (2/7)	33	37	1	0.11	0.13	1.1	99	598
Ship (2-8/7)	13	16	1	0.04	0.06	1.2	270	-
Ship (21-24/7)	8	8	0	0.04	0,04	1.0	-	-

Stationary T-PODs

The first study with stationary T-PODs off the coast of Sylt showed that harbour porpoises often swim by. All five T-PODs registered harbour porpoises on more than 90 % of the days in use (13 - 20 days in October 2002).

The second, longer deployment with three T-PODs from February through December 2003 confirmed this observation, with porpoises present on 98 - 99.5 % porpoise-positive days (Table 4). The longest deployment at one station was 288 days, with porpoises registered on 283 days (Table 4). The longest continuous logging of one T-POD was 70 days, with 69 porpoises-positive days.

Table 4. Overview of T-POD deployment at three stations (0.7 km, 1.2 km and 1.7 km from the coastline) in 2003 with porpoise-positive days and percentage of porpoise-positive days

	T-POD no.	Start date	End date	Deployed days	Porpoise pos. days	Percentage
Position 1 near to shore	158	26/2	Lost			
	170	5/5	17/6	44	43	97.7
	170	17/6	8/8	54	54	100
	170	8/8	23/9	46	45	97.8
	139	2/10	8/10	7	6	85.7
	170	6/11	Lost			
	Total			151	148	98
Position 2 center	170	26/2	24/4	58	57	98.3
	146	24/4	17/6	55	54	98.2
	147	17/6	8/8	53	53	100
	147	8/8	2/10	55	55	100
	268	2/10	6/11	35	34	97.1
	147	25/11	27/12	32	30	93.8
	Total			288	283	98.3
Position 3 away from shore	147	26/2	5/5	70	69	98,6
	127	5/5	17/6	44	44	100
	127	17/6	8/8	53	53	100
	146	8/8	28/8	20	20	100
	127	2/10	25/10	23	23	100
	146	6/11	Lost			
	Total			210	209	99.5

There was no significant difference in the daily presence (harbour porpoise-positive hours) of harbour porpoises between the three T-POD stations, at 0.7, 1.2 and 1.7 kilometres from shore, not in harbour porpoise
occurrence between ebb and flood tide was observed (Fig. 10). Porpoises
were frequently abundant during ebb and flood tide in the area.

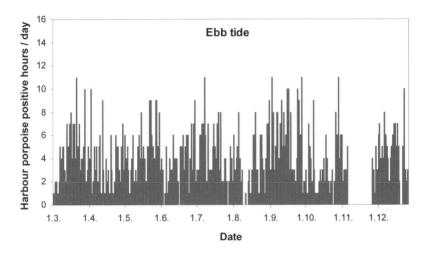

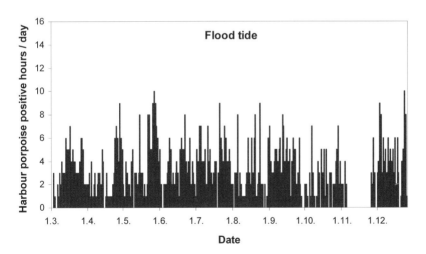

Fig. 10. Harbour porpoise-positive hours per day during ebb and flood tide at
POD-station no. 2 (1.2 km distance from shore). There were no data between 6th
and 25th of November due to device error

Towed T-POD

The sensor on the T-POD showed that the wings held the T-POD in a stable position 3.7 ± 0.19 m below the surface, and that the T-POD did not rotate vertically more than $11.2°$. The T-POD was towed during both of the surveys previously mentioned (August 2002 and July 2003). Two click trains were logged in July 2003, but both occurred at times when the ship was off the transect line and the observers were not active ("off effort"). The acoustically registered porpoises were therefore not spotted at the surface. The porpoise visually spotted nearest to the ship was 50 m away, but was not registered on the T-POD. No comparison between towed T-POD and surface observations could be made.

6.3.4   Impact of Offshore Windmills

The auditory evoked potentials were clearly identifiable and could be replicated repeatedly at varying frequencies and intensities. Also, the data from the captive harbour seal were comparable to the data measured with the free-ranging harbour seal.

However, the resulting hearing curve for the harbour seal (Fig. 11) only partially complies with the expected shape of the curve. Only at higher frequencies (≥ 8 kHz) do the threshold value reach lower values than for low frequencies. The hearing curve for marine mammals is typically U-shaped, i.e. the sensitivity increases steadily with increasing frequency to a maximum, before decreasing consistently (Richardson et al. 1995). The resulting AEP (Auditory Evoked Potentials) data for the animals tested in this study remain at a high threshold level (i.e. low sensitivity) for the low frequencies (0.125 kHz) up to a mid-frequency range (5.6 kHz), before they begin to sink to lower levels. In addition, sensitivity varied strongly between 0.7 and 5.6 kHz (max. ~20 dB). Systematic mistakes can be excluded as an explanation for the shape of the hearing curve, because the particular frequencies were measured in a random sequence. Masking can also be excluded because the measurements were conducted in a low-noise environment, and halted upon a loud noise. Moreover, the resulting threshold values represent an average of several thousand measurements collected over a period of minutes, thus excluding any short interference as a potential explanation. The shape of the hearing curve and the variations are very likely due to natural causes, und reflect impaired hearing sensitivity of the seal in the mid-frequency range (1 to 5.6 kHz).

Various acoustic stimuli were used for the harbour porpoise to measure its auditory threshold at 2 kHz. Nevertheless, the resulting values are in good accordance with each other (Fig. 12). The overall shape of the resulting hearing curve shows a close resemblance to the typical marine mammal hearing curve in this frequency range. In general, however, threshold values obtained with different types of acoustic stimuli showed a different degree of variation, i.e. the amplitude-modulated stimuli resulted in a more scattered distribution of the threshold values. The overall acoustic sensitivity of the harbour porpoise was approx. 10 - 20 dB higher than values gathered in behavioural studies by Andersen (1970) and Kastelein et al. (2002). This difference may be attributed to individual variability. However, the values measured in this study are below the corresponding values obtained by Popov and Supin in their 1990 ABR study. Individual variability may be an explanation, although it is more likely that a systematic difference exists due to the different methods used, i.e. ABR vs. the behavioural method.

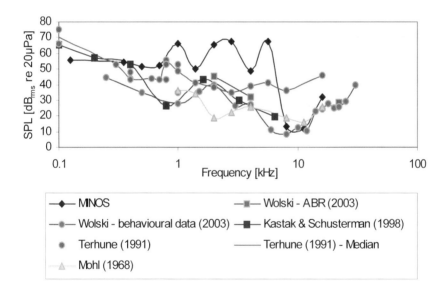

Fig. 11. Absolute AEP-hearing curve (blue line) of a harbour seal (*Phoca vitulina*) for sinusoidal signals in air and results from comparable studies (both AEP and behavioural studies). The sound pressure level is given on the ordinate, and the frequency on the abscissa

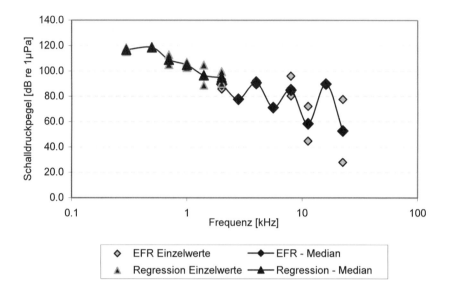

Fig. 12. AEP threshold values of a harbour porpoise (*Phocoena phocoena*) for sinusoidal signals at frequencies between 0.3 and 2 kHz as well as for amplitude-modulated signals at frequencies between 2 and 22.4 kHz. The curve runs along the median values at the particular frequencies. The sound pressure level is given on the ordinate, and the frequency on the abscissa

6.4   Conclusion

6.4.1   Density, Distribution Patterns and Habitat Use in North and Baltic Sea

Aerial surveys revealed the presence of harbour porpoises throughout the German North Sea with a gradient from north to south for the summer months of 2002 and 2003. The abundance estimates for the North Sea were stable over the years. The area with the highest density was found to be west of the island of Sylt, the so-called 'Amrum Aussengrund'. T-POD data obtained close to the shore of Sylt confirmed the high harbour porpoise density. Stationary T-PODs registered porpoises almost daily in 2003, regardless of the season. The influence of tidal changes or diurnal differences on the usage of the area by harbour porpoises could be investigated with the devices. The investigations revealed no influence due to tide or time of day on the presence of porpoises.

In the Baltic Sea, the mean abundance of harbour porpoises during the summer months of 2002 and 2003 was considerably lower than in the North Sea. The sighting rate differed considerably from year to year, particularly in area G, the Pomeranian Bight. Here, sightings of porpoises revealed a temporary high abundance of porpoises in June and July 2002, possibly caused by the presence of local food sources or social aggregations, e.g. for reproduction. Nonetheless, data analysed for both years showed a decrease in mean abundance in the Baltic Sea from west to east in the summer months of 2002 and 2003. This was confirmed by the T-POD data. Unfortunately, no T-PODs were deployed in the Pomeranian Bight at the time of the high rate of porpoise sightings in June and July 2002. Only very few porpoise registrations were obtained with T-PODs afterwards, confirming the low density observed in aerial surveys.

Morphological and genetic studies revealed the existence of a separate subpopulation of harbour porpoises in the Baltic proper, i.e. east of the Darss and Limhamn underwater ridge (Huggenberger et al. 2002, Tiedemann et al. 2001). The low density of this subpopulation confirmed by the present studies raises deep concern for the survival of the population, as stressed in the Recovery plan for Baltic Harbour Porpoises (Jastarnia Plan, ASCOBANS). Any negative anthropogenic influence (e.g. fishery bycatch, chemical or noise pollution) on this very small and therefore highly endangered subpopulation may sooner or later lead to its extinction if no action is taken.

The long-term deployment of T-PODs yielded a seasonal picture of the habitat use of porpoises in the Baltic Sea. Seasonal changes around Fehmarn (area I) and in the Kadet Channel and adjacent coastal waters (area II) were discovered, in addition to the pronounced decrease in the percentage of porpoise-positive days per month from the western part of the German Baltic Sea around the island of Fehmarn to the eastern part up to the Pomeranian Bight. Verfuß et al. (2005) showed the importance of echolocation for harbour porpoises. Porpoises living in a well-known, semi-natural outdoor pool used echo-location at all times, even for simple orientation tasks during daylight, regardless of the season. Regular use of echolocation by harbour porpoises is therefore very likely. The variation in the amount of porpoise registrations throughout the year and differences across areas were presumably caused by temporal changes and geographical differences in harbour porpoise density. These changes were proven not to be affected by the different sensitivities of the T-PODs used (Verfuß et al. 2006).

Until the mid-20th century, migration of harbour porpoises was assumed for the North and Baltic Sea (reviewed in Koschinski 2003). In spring, the porpoises were thought to follow the seasonal movements of herring, passing Danish waters into the Baltic Sea. In late autumn and winter, when the Baltic Sea froze over in some years, the porpoises may have migrated back out of the Baltic Sea. Nowadays, the porpoise stocks are too small to easily verify such migrations. Teilmann et al. (2004) could prove seasonality in Danish waters with the help of satellite tags on porpoises. Siebert et al. (accepted) showed seasonality in incidental sightings and stranding rates in the German Baltic Sea, with a peak in the summer months. The data from incidental sightings may be biased by a lower effort in winter (e.g. fewer sailing boats), whereas stranding events can be biased by a longer submersion time of carcasses when water temperature is low (Moreno 1993, in Siebert et al., accepted) The T-PODs proved seasonal changes in the use of the Baltic Sea areas around Fehmarn and the Kadet Channel.

6.4.2   Intercalibration

T-PODs in stationary deployment proved to be valuable tools for investigating the habitat use of harbour porpoises and to reveal temporal and/or geographical differences in specific areas, but towed T-PODs proved cumbersome. Their shape and directional receiving pattern may not be optimised for this kind of application. A towed hydrophone array will be tested in further investigations.

For comparision of visual and acoustic survey methods, more data are needed to estimate g(0) and allow for direct statistical comparison.

6.4.3   Impact of Offshore Windmills

The results of this study confirm that the chosen Auditory Brainstem Response (ABR) method is suitable for the study of the absolute hearing threshold of marine mammals and were successfully established within this project. The method can be applied to measurements under controlled conditions as well as in the wild. Auditory measurements are feasible in air and under water. A full in-air audiogram of a harbour seal and a partial audiogram of a harbour porpoise were obtained. The harbour porpoise audiogram requires further measurements above 22.4 kHz, and more animals of both species must be tested. The ABR method may be used for such further measurements on captive animals to improve the declarative strength of the baseline data. ABR measurements may thus become a tool for ecological survey programs with wild-caught animals, if more experience is gained regarding the precise assessment of auditory thresholds under suboptimal conditions.

References

Andersen S (1970) Auditory Sensitivity of the Harbour Porpoise (*Phocoena phocoena*). Inv in Cet 2:255-259

Au WWL (1993) The sonar of dolphins. Springer-Verlag, New York Berlin Heidelberg

Benke H, Siebert U (1994) Zur Situation der Kleinwale im Wattenmeer und in der südlichen Nordsee. In: Lozán JL, Rachor E, Reise K, v. Westernhagen H, Lenz W (eds). Warnsignale aus dem Wattenmeer. Blackwell Wissenschaftsverlag, Berlin, pp 309-316

Benke H, Siebert U, Lick R, Bandomir B, Weiss R (1998) The current status of harbour porpoises (*Phocoena phocoena*) in German waters. Arch Fish Mar Res 46(2):97-123

Berggren P (1995) A preliminary assessment of the status of harbour porpoises (*Phocoena phocoena*) in the Swedish Skagerrak, Kattegat and Baltic seas. Paper SC/47/SM50 presented to the IWC Scientific Committee, May 1995, 22 pp

Börjesson P, Berggren P (1997) Morphometric comparisons of skulls of harbour popoises (*Phocoena phocoena*) from the Baltic, Kattegat and Skagerrak seas. Can Jour Zool 75:280-287

Buckland ST, Anderson DR, Burnham KP, Laake JL, Borchers DL, Thomas L (2001) Introduction to distance sampling. Estimating abundance of biological populations. Oxford University Press, Oxford, 432 pp

Caretta JV, Taylor BL, Chivers SJ (2001) Abundance and depth distribution of harbor porpoise (*Phocoena phocoena*) in northern California determined from a 1995 ship survey. Fish Bull 99:29-39

Goodson AD, Kastelein RA, Sturtivant CR (1995) Source levels and echolocation signal characteristics of juvenile harbour porpoises (*Phocoena phocoena*) in a pool. In: Nachtigall PE, Lien J, Au WWL, Read AJ (eds) Harbour porpoises - laboratory studies to reduce bycatch. Vol 1. De Spil Publishers, Woerden, The Netherlands, pp 41-53

Hammond PS, Benke H, Berggren P, Borchers DL, Buckland ST, Collet A, Heide-Jørgensen MP, Heimlich-Boran S, Hiby AR, Leopold MP, Øien N (1995) Distribution and abundance of the harbour porpoise and other small cetaceans in the North Sea and adjacent waters. Final Report, LIFE 92-2/UK/027, 242 pp

Hammond PS, Berggren P, Benke H, Borchers DL, Collet A, Heide-Jørgensen MP, Heimlich-Boran S, Hiby AR, Leopold MF, Øien N (2002) Abundance of harbour porpoise and other cetaceans in the North Sea and adjacent waters. J of Appl Ecol 39:361-376

Hiby AR, Hammond PS (1989) Survey techniques for estimating the abundance of cetaceans. Reports of the International Whaling Commission, Special Issue 11:47-80

Huggenberger S, Benke H, Kinze CC (2002) Geographical variation in harbour porpoise (*Phocoena phocoena*) skulls: support for a separate non-migratory population in the Baltic Proper. Ophelia 56 (1):1-12

Hutchinson J, Simmonds M, Moscrop A (1995) The harbour porpoise in the North Atlantic: a case for conservation. Conservation Research Group, University of Greenwich. London, January 1995. Report to Stitching Greenpeace Council: 90 pp

IWC (1997) Report of the Sub-Committee on Small Cetaceans, Annex H. Report of the International Whaling Commission 47:169-191

Jepson PD, Baker JR, Allchin CR, Law RJ, Kuiken T, Baker JR, Rogan E, Kirkwood JK (1999) Investigating potential associations between chronic exposure to polychlorinated biphenyls and infectious disease mortality in harbour porpoises from England and Wales. Sci Tot Environ 243/244:339-348

Kamminga C, Engelsma FJ, Terry RP (1999) An adult-like sonar wave shape from a re-habilitated orphaned harbour porpoise (*Phocoena phocoena*). Ophelia 50:35-42

Kaschner K (2001) Harbour porpoises in the North Sea and Baltic – bycatch and current status. Report for the Umweltstiftung WWF – Deutschland: 82 pp

Kastelein RA, Bunskoek P, Hagedoorn M, Au WWL (2002) Audiogram of a harbour porpoise (*Phocoena phocoena*) measured with narrow-band frequency-modulated signals. J Acoust Soc Am 112(1):334-344

Kinze CC (1995) Exploitation of harbour porpoises (*Phocoena phocoena*) in Danish waters: a historical review. In: Bjorge A, Donovan GP (eds) Biology of the Phocoenids. Black Bear Press, Cambridge, pp 141-153

Kock KH, Benke H (1996) On the by-catch of harbour porpoise (*Phocoena phocoena*) in German fisheries in the Baltic and the North Sea. Arch Fish Mar Res 44(1/2):95-114

Koschinski S (2003) Current knowledge on harbour porpoises (*Phocoena phocoena*) in the Baltic Sea. Ophelia 55:167-197

Lockyer C, Kinze C (2000) Status and life history of harbour porpoise, *Phocoena phocoena*, in Danish waters. ICES, Copenhagen: pp 1-37

Moreno P, Benke H, Lutter S (1993) Behaviour of harbour porpoise (*Phocoena phocoena*) carcasses in the German Bight: surfacing rate, decomposition and drift routes. In: Bohlken, H, Benke, H, Wulf, J (eds) Untersuchungen über Bestand, Gesundheitszustand und Wanderung der Kleinwalpopulationen (*Cetacea*) in deutschen Gewässern. FKZ 10805017/11. BMU-Final Report. Institut für Haustierkunde, University of Kiel, Germany

Ødegaard and Danneskiold-Samsøe A/S (2000) Offshore Wind-Turbine Construction. Offshore Pile-Driving Underwater and Above-Water Noise Measurements and Analysis. SEAS Distribution A.m.b.A. und Enron Wind GmbH, Report no. 00.877, Denmark.

Popov VV, Supin AY (1990b) Electrophysiological studies of hearing in some cetaceans and manatee. pp 405-415. In: Thomas JA and Kastelein RA (eds): Sensory Abilities of Cetaceans: Laboratory and Field Evidence. Plenum Press, New York, USA

Reijnders PJH (1992) Harbour porpoises *Phocoena phocoena* in the North Sea: Numerical responses to changes in environmental conditions. Netherlands J. Aqua Ecol 26(1):75-85

Richardson WJ, Greene CR, Jr, Malme CI, Thomson DH (1995) Marine Mammals and Noise. Academic Press, San Diego, USA

Scheidat M, Siebert U (2003) Aktueller Wissensstand zur Bewertung von anthropogenen Einflüssen auf Schweinswale in der deutschen Nordsee. Seevögel 24(3)50-60

Scheidat M, Gilles A, Siebert U (2004) Teilprojekt 2 – Erfassung der Dichte und Verteilungsmuster von Schweinswalen (*Phocoena phocoena*) in der deutschen Nord- und Ostsee. In: Kellermann A et al. (eds) Marine Warmblüter in Nord- und Ostsee: Grundlagen zur Bewertung von Windkraftanlagen im Offshore-Bereich. FKZ 0327520. Final report. Investment-in-future program of the German Federal Ministry for the Environment, Nature Conservation and Nuclear Safety (BMU). Landesamt für den Nationalpark Schleswig-Holsteinisches Wattenmeer, Tönning. (www.wattenmeer-nationalpark.de)

Schulze G (1996) Die Schweinswale. Neue Brehm Bücherei, Magdeburg, 188 pp

Siebert U, Benke H, Schulze G, Sonntag RP (1996) Über den Zustand der Kleinwale. In: Lozán JL, Lampe R, Matthäus W, Rachor E, Rumohr H, von Westernhagen H (eds) Warnsignale aus der Ostsee. Parey Buchverlag, Berlin, pp 242 - 248

Siebert U, Joiris C, Holsbeek L, Benke H, Failing K, Frese K, Petzinger E (1999) Potential relation between mercury concentrations and necropsy findings in cetaceans from German waters of the North and Baltic Seas. Mar Poll Bull 38:285-295

Siebert U, Gilles A, Lucke K, Ludwig M, Benke H, Kock K-H, Scheidat M (accepted) Review of occurrence of the harbour porpoise (*Phocoena phocoena*) in German waters – analyses of aerial surveys, incidental sightings and strandings. J Sea Res

Teilmann J, Dietz R, Larsen F, Desportes G, Geertsen BM, Andersen LW, Aastrup PJ, Hansen JR, Buholzer L (2004) Satellitsporing af marsvin i danske og tilstødende farvande. Danmarks Miljøundersøgelser. Report. DMU 484, pp 1-86. Electronic version: www2.dmu.dk/1_viden/2_Publikationer/3_fagrapporter/abstrakter/abs_484DK.asp

Tiedemann R (2001) Stock definition in continuously distributed species using molecular markers and spatial autocorrelation analysis. SC/53/SD3, pp 1-4. London. Paper presented to the Scientific Committee of the International Whaling Commission

Tiedemann R, Harder J, Gmeiner C, Haase E (1996) Mitochondrial DNA patterns of harbour porpoises (*Phocoena phocoena*) from the North and the Baltic Sea. Zeitschrift für Säugetierkunde 61:104-111

Verfuß UK, Schnitzler H-U (2002) Untersuchungen zum Echoortungsverhalten der Schweinswale (*Phocoena phocoena*) als Grundlage für Schutzmaßnahmen. R+D project: FKZ: 898 86 021, Final report, pp 1-53. Eberhard Karls University of Tübingen, Dept Animal Physiology, Germany

Verfuß UK, Honnef C, Benke H (2004a) Untersuchungen zur Nutzung ausgewähl-
    ter Gebiete der Deutschen und Polnischen Ostsee durch Schweinswale mit
    Hilfe akustischer Methoden. R+D project: FKZ: 901 86 020. Final report,
    German Oceanographic Museum, Stralsund, Germany
Verfuß UK, Honnef C, Benke H (2004b) Teilprojekt 3 – Untersuchungen zur
    Raumnutzung durch Schweinswale in der Nord- und Ostsee mit Hilfe akusti-
    scher Methoden (PODs). In: Kellermann, A et al. (eds) Marine Warmblüter in
    Nord- und Ostsee: Grundlagen zur Bewertung von Windkraftanlagen im Off-
    shore-Bereich. FKZ 0327520. Final report. Investment-in-future program of
    the German Federal Ministry for the Environment, Nature Conservation and
    Nuclear Safety (BMU). Landesamt für den Nationalpark Schleswig-
    Holsteinisches Wattenmeer, Tönning. (www.wattenmeer-nationalpark.de)
Verfuß UK, Miller LA, Schnitzler H-U (2005) Spatial orientation in echolocating
    harbour porpoises (Phocoena phocoena). J Exp Biol 208:3385-3394
Verfuß UK, Honnef CG, Benke H (2006) Seasonal and geographical variation of
    harbour porpoise habitat use in the German Baltic Sea monitored with auto-
    nomous echolocation click detectors (T-PODs). In: von Nordheim H, Boede-
    ker D, Krause JC (eds) Advancing towards effective marine conservation in
    Europe – NATURA 2000 sites in German offshore waters. Springer, Heidel-
    berg
Vinther M (1999) Bycatches of harbour porpoises (Phocoena phocoena L.) in
    Danish set-net fisheries. J Cet Res Manag 1(2):123-135

7   Distribution of Harbour Seals in the German Bight in Relation to Offshore Wind Power Plants

Dieter Adelung, Mandy A M Kierspel, Nikolai Liebsch, Gabriele Müller, Rory P Wilson

7.1   Introduction

Harbour seals, *Phoca vitulina*, are the most numerous of the two seal species of the Wadden Sea. They use the sand banks for resting, moulting, pupping and lactation during low tides. Since 1974, when hunting was strictly forbidden by the governments of The Netherlands, Germany and Denmark, the population increased continuously from 5,400 animals to around 15,000 seals in 1988, when an epizootic caused by the so called phocine distemper virus reduced the stock by approximately 60 % (Kennedy 1990; Reijnders et al. 1997). The population recovered faster then expected, and by 2002 reached a level of around 29,000 seals. Then another epizootic occurred again, induced by the same or a similar virus to that of 1988. About 50 % of the population died this time (Reijnders et al. 2003c). As of 2005, the population had increased to about 20,000 harbour seals in the Wadden Sea. Given the rate of recovery so far, it is expected that by 2007 the population will reach the pre-outbreak size, provided that the seals operate under the same conditions as before and without new impacts (Reijnders et al. 2005). Possible impacts include reduced food supplies due to increased pressure from fisheries, and disturbance by the construction and operation of offshore wind farms (assuming that such power plants are built as planned).

All offshore wind farms are scheduled to be built outside the Wadden Sea. Initially therefore, no conflict between seals and wind farms is expected, as the seal haul-out sites are outside the affected area. However, seals spend around 80 % of their time in the water, in and outside the Wadden Sea. If offshore wind farms are built in the proposed areas and these prove to be important for the seals, conflicts are possible. But what kind of impact by windmills is expected?

The main impact will probably be observed during the construction phase, from the noise of ramming and the resulting increased turbidity of the water, which may hinder seals and also their prey.

During operation of these parks, the noise of rotating propellers emitted into the water could repel both seals and fish. An additional problem may occur when a series of wind farms is constructed, which could block migratory routes of seals from their haul-outs to foraging areas. Currently, there is little information as to how windmills might affect seals. The most relevant study was done by Tougaard et al. 2003. They deployed satellite transmitters on seals before and after construction of the offshore wind farm in Horns Reef in the Danish part of the North Sea. The data indicate that there was no effect on seals during construction or operation. But satellite transmitters provide very few indications as to the route of seals and data on their behaviour and activity are rare.

Conversely, wind farms could also be beneficial to seals, since fishing would be banned in these areas and new benthic organisms could settle there using the hard substrate of the pylons within this otherwise sandy area. This could attract fish, creating a refuge area.

In order to ascertain whether there are impacts upon or benefits to the seals, it is necessary to learn more about their movements and activities.

Although radio- and satellite telemetry can ascertain useful data on seal habitat use at sea (e.g. Weimerskirch et al. 1993), transmissions can only occur if the antenna is out of the water, so that data on diving activities are limited (McConnell et al. 2004). For this reason, we have developed, together with the company Driesen and Kern (Bad Bramstedt, Germany), a satellite supported "dead-reckoning-system" (Wilson and Wilson 1988; Mitani et al. 2003), which provides continuous records of all important activities of the seals on land and in the water for periods of up to three months.

7.2   Methods

The data logger records information from ten to twelve different sensors every 3 to 20 seconds which are stored on a chip with a memory capacity of up to 32 MB. It stores information on heading from a three-dimensional compass, swim speed, dive depth, water temperature, light intensity, body orientation, pitch, and roll. From these data it is possible to reconstruct all movements and activities of the seals at sea and on land, without any breaks.

The main disadvantage of this method is that the devices have to be retrieved to access the stored data. For technical reasons, it is impossible to re-catch the equipped seals to remove the devices, so that an ARGOS satellite tag (PTT) and a timer mechanism for automatic release of the unit are placed together in a pressure-resistant hull of buoyant material.

The total unit has a weight of approx. 670 g, a length of 180 mm, a width of 90 mm, and a height of 60 mm. The weight of the seals was on average 77.2 kg, which means that the devices in total were less than 1 % of body weight on land (0.87 %), whereas at sea they were slightly positively buoyant.

Depending on their haul-out site, the seals where caught with a hundred metre long net coming from seaside at Lorenzenplate and Rømø (Orthmann 2000), or on Helgoland by surprising and catching them while resting on the beach. The seals there are very familiar with tourists, and ignore them up to a distance of about ten metres.

After capture, the seals were immediately transferred individually into ring nets and cooled with seawater from time to time to minimize heat stress. To attach the device in the middle of the back, the seals were then strapped to a bench with belts and, after drying the fur, a base of neoprene containing the logger unit was glued to the fur with fast-setting epoxy glue (Fig. 1).

Fig. 1. Seal equipped with device (coloured in red) entering the water

The unit containing the devices is programmed to be released at a preset time, and is usually washed ashore after a while by the currents and the prevailing westerly winds. Once ashore, the unit can be located by the PTT, or is found by beach walkers. The probability of finding the unit in this region is close to 70 %. The neoprene base comes off during the seals' annual moult.

The PTT provides information not only on where a unit could be found, but also of the time and location of haul-out events of the seals. PTT positional data at sea are very rare, since the logger is on the seals' back, and does not usually emerge when the animals come up for air.

After retrieval, the stored data can be downloaded and analysed by special software, which permits calculation of the routes taken by the seals and displays their dive profiles during various activities such as travelling, searching, foraging or resting in the water as well as on land.

The devices can be re-used after the batteries are replaced.

7.3   Results

This report includes the results of the MINOS TP6[1] project, which was concluded in February 2004 after three years as well as the first results of the project MINOS⁺ TP 6, which began in June 2004.

7.3.1   MINOS

During the MINOS project, 21 seals were equipped with loggers and 17 devices were retrieved. Data could be downloaded from 13 of these loggers, corresponding to 153 seal days. Of the 21 devices, 19 devices were deployed on seals from the Lorenzenplate sandbank, situated in the neighbourhood of the Eiderstedt Peninsula in Schleswig-Holstein. The Lorenzenplate is a major haul-out site for up to several hundred harbour seals.

Two seals were also equipped at the beach of Rømø Island in Denmark. This haul-out site is also regularly visited by many seals.

Eight devices were installed on females. Of the five devices installed in the spring, four returned. One of these did not operate correctly, two only for a few days and only one for a longer period. Hence, the data do not permit conclusions on seasonal or sex differences.

---

[1] MINOS TP 6 - MINOS sub-project 6: Telemetry and seals

7.3.2   MINOS⁺

During the current project (MINOS⁺) a total of 41 seals have already been equipped with loggers at three different sites: at the Lorenzenplate (17), on Rømø (19) and on Helgoland (5). The distribution of the devices by location, season, and sex are shown in Table 1.

Table 1. Distribution of devices deployed in the MINOS⁺ project

		Spring	Autumn	Winter
Lorenzenplate	Females	3 (1)	1 (1)	3 (3)
	Males	5 (4)	2 (1)	3 (1)
Rømø	Females	-	-	-
	Males	11 (5)	6 (5)	2 (0)
Helgoland	Females	2 (2)	-	-
	Males	-	3 (2)	-

The values shown represent the numbers of seals equipped, and recovered, (in brackets) respectively.

To date, 25 of these devices have been recovered. Data from 19 devices have been downloaded, resulting in 626 days of dive data and 101 shorter or longer trips at sea. With an average of 260 dives per day, this results in more than 160,000 dives. Most of the devices have been recovered recently, so that the dive analysis is still pending. Data from four devices have yet to be downloaded.

7.3.3   Foraging Areas

The data show that all seals left the Wadden Sea and moved to deeper offshore areas, where foraging activity dominated their behaviour. The animals left their resting areas (sandbanks or beaches) and headed more or less straight to distinct offshore areas, thereby reaching distances from land of approximately 100 km (Fig. 2). Most foraging trips were between one day and two weeks, with a mean of five to six days. However, during very short trips, the seals might not have enough time to reach deeper water and thus are unlikely to be feeding extensively during these trips. Such short trips are often found between trips of longer duration, and may be caused by tidal flooding of their resting areas.

Seals from both the Lorenzenplate and Rømø show the same general pattern of movements (Fig. 2). They start out travelling in a westerly direction, heading for deeper waters. Once they have reached their foraging grounds, they move in a less directed manner, on highly winding routes, spending a significant amount of time in specific areas. On their way back to the resting areas, they again move in a more direct path.

Of particular interest is the foraging behaviour of two seals from Helgoland. These animals performed foraging trips of less than a day each for a period of four days and of about six weeks, respectively. This suggests that the animals stayed around the island of Helgoland during the entire time the device functioned, which was confirmed by the PTT data. They could apparently reach suitable foraging grounds within a few hours.

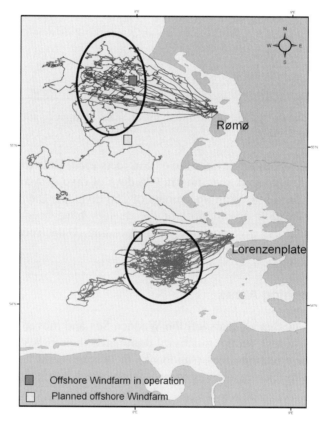

Fig. 2. Foraging routes of harbour seals from the Lorenzenplate (based on three animals) and from Rømø (based on two animals)

7.3.4 Diving Behaviour

To date, very little of the available dive data from MINOS[+] has been sub-jected to the extensive and time-consuming analysis. However, the existing information accords with results from the first MINOS project.

Generally, harbour seals perform U-shaped dives, which indicate that they make directed descents to the sea bed, move along the bottom, and ascend directly to the surface (Fig. 3). The mean duration of these dives is approx. five minutes. In shallow water, mainly in the Wadden Sea area, they also make shorter dives, with the descent followed directly by the ascent, thus creating a V-shaped dive profile. A view of entire trips reveals that dive depth follows the bathymetry. It thus appears that the seals nearly always dive to the bottom.

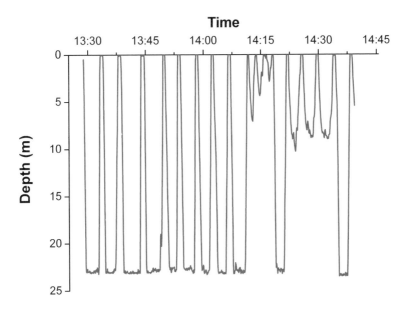

Fig. 3. Example of dive profiles of a harbour seal, showing a series of U-shaped dives followed by several V-shaped dives

Based on tilt information, the U-shaped dives can break down into "active" (high changes of pitch and roll signals) and "passive" (low

changes of pitch and roll signals) dives. Active dives involve such behaviour as travelling and foraging, while passive dives are resting dives, both in shallow and in deep water. Periods of up to one hour including resting dives are particularly apparent during longer foraging trips. Seals from the Lorenzenplate sandbank are only able to haul out during low tide.
The high tide is spent in the water, where they sometimes also perform long resting dives (Adelung et al. 2004).

7.4   Discussion

Due to the fact that the devices are deployed on the back of the seals, no good quality locations are ascertained through the satellite transmitters from animals at sea. Nearly all signals are therefore related to resting activity of the seals on the sandbanks. Resting around the sandbanks or at the beach of Rømø lasts several days, interrupted only for several hours by the daily tides. These resting periods are then followed by foraging trips, which last up to 12 days, after which the seals return to their haul-out sites. By contrast, the first data sets from seals from Helgoland indicate quite a different strategy. The satellite data show a daily presence of the seals at the haul-out site. Accordingly, the logger data from the two seals show that they perform only short foraging trips, lasting less than 24 hours. The main reason for this different behaviour relates to their haul-out site, the island of Helgoland, which is located in the open sea, and apparently closer to their foraging areas.

Thus, Helgoland animals exhibit foraging trips with almost no transit time, which makes short trips more efficient, whereas seals from the Wadden Sea engage in longer trips to compensate for the effective loss of energy incurred during the transit phases of their trips, during which only little food can be found.

For all these foraging trips, the tracks show clearly that the seals leave the coastal region and head for deeper offshore waters, where the abundance of prey is presumed higher.

Based on this information, it can be concluded that the areas of planned and operating offshore wind farms lie inside the foraging grounds of the harbour seals of the German Bight, and that there is therefore a potential for conflict, although its scope and nature is unclear.

The tracks of "Danish" seals show no obvious reaction to the operating wind park in this area, but as these observations only started after the plant started producing, and no data with similar temporal and spatial resolution are available, no comparison with the "pre-wind park-period" could be made.

What has to be kept in mind is that this research only alludes to possible changes in behavioural patterns, but can give no information on other cues, which might be of various origins (e.g. acoustic stress). Moreover, even if the direct impact on the seals turns out to be of minor importance, factors influencing seal prey (mainly benthic fish) could become relevant. Such impacts are not necessarily negative. As the North Sea has a largely sandy bottom, a hard substrate, such as the foundations of wind generators, can act as a kind of artificial reef, creating a new ecological niche, and attracting a variety of organisms which could include fish, and subsequently seals.

In summary, this study has been conducted over several years, and provides high resolution baseline data on the behaviour of harbour seals in this region. It gives an insight into various activity patterns, and where they occur. Possible changes can be elucidated, but we still know little about the underlying reasons. Further studies on the sensory systems and data describing the environment used by the animals (e.g. oceanographic parameters, noise, etc.) are essential for a comprehensive picture of harbour seal ecology and in particular, for an elucidation as to how the animals might react to changed conditions.

7.5   Summary

Determination of how animals allocate activities (such as resting, feeding and breeding) to space and time is fundamental for understanding their ecology, their role in ecosystems, and in assessing the impact of their environment on them (e.g. as possibly caused by offshore wind farms).

In order to obtain such data in the context of the MINOS[2] and MINOS[+] project, we equipped harbour seals in the German Bight with dead-reckoners and satellite transmitters to track them while they were hauled out on land and foraging at sea. Typical movement patterns are presented, and possible influences of offshore wind farms are discussed.

---

[2] MINOS - Marine Warmblüter in Nord- und Ostsee (Marine warm-blooded animals in the North and Baltic Seas: Foundations for assessment of offshore wind farms)

References

Adelung D, Wilson RP, Liebsch N, (2004) MINOS – Marine warm-blooded animals in the North and Baltic Seas: Foundations for assessment of offshore wind farms. Summary on subproject 6 – harbour seal telemetry.
http://www.wattenmeer-nationalpark.de/themen/minos/minos2_download_rep.htm

Kennedy S (1990) A review of the 1988 European seal epizootic. Vet Rec 1990, 563-567

McConnell B, Beaton R, Bryant E, Hunter C, Lovell P, Hall A (2004) Phoning home – A new GSM mobile phone telemetry system to collect mark-recapture data. Marine Mammal Science, 20 (2):274-283

Mitani Y, Sato K, Ito S, Cameron MF, Siniff DB, Naito Y (2003) A method for reconstructing three-dimensional dive profiles of marine animals using geomagnetic intensity data: results from two lactating Weddell seals. Polar Biol. 26:311-317

Orthmann T (2000) Telemetrische Untersuchungen zur Verbreitung, zum Tauchverhalten und zur Tauchphysiologie von Seehunden (*Phoca vitulina vitulina*) des Schleswig-Holsteinischen Wattenmeeres. PhD Dissertation. Christian-Albrechts-University of Kiel. Germany

Reijnders PJH, Ries EH, Tougaard S, Nørgaard N, Heidemann G, Schwarz J, Vareschi E, Traut IM (1997) Population development of harbour seals *Phoca vitulina* in the Wadden Sea after the 1988 virus epizootic. J Sea Res 38:161-168

Reijnders PH, Brasseur SMJM, Brinkman AG (2003c) The phocine distemper virus outbreak of 2002 amongst harbour seals in the North Sea and the Baltic Sea: spatial and temporal development and predicted population consequences. In: Management of North Sea harbour and grey seal population. Proceedings of the International Symposium at Eco-Mare, Texel, The Netherlands, November 29-30, 2002. Wadden Sea Ecosystem No. 17, Wadden Sea Secretariat, Wilhelmshaven, Germany. pp 19-25

Reijnders PJH, Abt K, Brasseur SMJM, Camphuysen KCJ, Reineking B, Scheidat M, Siebert U, Stede M, Tougaard J, Tougaard S (2005) Marine Mammals in Wadden Sea Ecosystem 19:305-318

Tougaard J, Ebbesen I, Tougaard S, Jensen T, Teilmann J (2003) Satellite tracking of harbour seals on Horns Reef. Technical report to Techwise A/S, Biological Papers from the Fisheries and Maritime Museum. Esbjerg. No. 3
Available at:
http://www.hornsrev.dk/Miljoeforhold/miljoerapporter/Hornsreef%20Seals%202002.pdf

Weimerskirch H, Salamolard M, Sarrazin F, Jouventin P (1993) Foraging strategy of wandering albatrosses through the breeding season: a study using satellite telemetry. Auk 100:325-342

Wilson RP, Wilson MP (1988) Dead Reckoning – a New Technique for Determining Penguin Movements at Sea. Meeresforschung – Reports on Marine Research 32:155-158

8   Research on Marine Mammals Summary and Discussion of Research Results

Barbara Frank

8.1   Introduction

The first application for an offshore wind farm in German waters was approved in 2001. It then became obvious that extensive research was required in order to properly assess the effects of wind farms on the marine environment prior to planning and construction of windmills. Until then, there was plainly not enough knowledge of marine mammals in German offshore waters to do so. The Environmental Impact Assessments (EIA) which the wind energy companies have to conduct provide only localised site-specific studies. These studies can only supply a partial picture, and do not enable population-level assessments. Since 2002, the large-area studies of the research projects MINOS[1] and MINOS[+] have covered the entire German Exclusive Economic Zone (EEZ). They have greatly enhanced the knowledge of seals, harbour porpoises and seabirds in our waters, and identified important habitat areas for these species. During these projects, considerable methodological development has also been undertaken, which will facilitate further research and monitoring. Nevertheless, despite all this progress, it is still very difficult to picture distinct impacts of offshore wind farms on marine mammals. Experience exists with small wind farms close to the Scandinavian coasts, but it cannot be transferred directly to the situation which pertains in German waters and planning sites. The first big wind farms were not brought into operation until December 2002, at Horns Reef, off the west coast of Denmark, and a year later in the shallow waters of the Danish Baltic, at Nysted/Rødsand. Since then, initial experience has been gained during operation.

---

[1]   MINOS - Marine Warmblüter in Nord- und Ostsee (Marine warm-blooded animals in the North and Baltic Seas: Foundations for assessment of offshore wind farms)

8.2   MINOS Results on Harbour Seals

The results obtained to date from MINOS and MINOS[+] have provided the first deep insights into the underwater life of our harbour seals. They have shown that their habitat is not restricted to the vicinity of their haul-out sites in the Wadden Sea. They appear to rely on foraging off the coast, where the abundance of prey is probably higher. Seals therefore cover long distances of up to 120 km, and undertake foraging trips lasting several days. They would presumably not do so if they could reach adequate feeding grounds within a few hours, as the comparison with the first data sets from seals from the island of Helgoland indicate (see chapter 7). Offshore foraging thus seems to be vital for the Wadden Sea seals. Future conflicts with the offshore wind farms could hence be possible, since the planning application boundaries partially overlap with the feeding grounds. The wind farms might also form a barrier to the animals' migratory routes (Tougaard et al. 2003a). However, Adelung et al. (chapter 7) point out that research to date is merely able to indicate possible changes in behavioural patterns. But it can hardly provide any evidence of the underlying reasons as e.g. permanent acoustic stress.

8.3   MINOS Results on Harbour Porpoises

8.3.1   German North Sea

In 2002, the EIA for the German Butendiek wind farm project in the North Sea reported the highest numbers of harbour porpoises and mother-calf pairs in German waters ever registered up to that date. The study covered an area of about 2,500 km^2 within the Outer Sylt Reef. The results of all ship-based and aerial surveys showed that the porpoises were widely distributed with high spatial variation (BioConsult SH and GFN 2002). The authors of the EIA concluded that impacts within the 40 km^2 of the planning site could be neglected with respect to the size of the entire survey area. The planning site seemed to be as well suited for a wind farm as the surrounding areas under investigation. It was only one year later that the first MINOS results showed that the high densities of porpoises recorded in the Butendiek study could not be assumed for the entire German North Sea (Siebert et al. 2003).

Similar numbers were in fact only found at the same Outer Sylt Reef. And although harbour porpoises were present throughout the German North Sea, there was a decrease from north to south. The MINOS data revealed that the densities of harbour porpoises on the Outer Sylt Reef were up to ten times higher than in the other parts of the German North Sea. The abundance estimates from the 2002 and 2003 summer surveys were the first ever to be published for the entire German North Sea. They proved to be stable during these two years, with 34,000 and 39,000 porpoises respectively (Siebert et al. 2004).

Moreover, the Butendiek EIA had observed a distinct seasonal pattern, with low numbers in winter and maximum densities in spring and summer (BioConsult SH and GFN 2002). This pattern has been observed in all subsequent observations in this region (chapter 6, Grünkorn et al. 2004, Diederichs et al. 2004). The currently ongoing MINOS[+] surveys have also confirmed the observations of the preceding years: The number of harbour porpoises at the Outer Sylt Reef starts to increase as of April and decreases again towards autumn.

8.3.2 German Baltic Sea

A quite different picture with much lower porpoise numbers was found for the German EEZ in the Baltic Sea. The mean abundances of the summer surveys taken in 2002 and 2003 differed, with 4,600 and 1,600 porpoises, respectively. The distribution in 2002 was for the most part different from that in 2003, since it included a hitherto singular and temporary hot spot in the Pomeranian Bight (chapter 6). This concentration was an important finding, as it revealed the potential spatial and temporal dynamics of harbour porpoise distributions, which are presumably due to rich and moving food sources. Moreover, it focused upon the local harbour porpoise population of the Baltic proper, which is assumed to be depleted and highly endangered (chapter 6).

Particularly via the use of T-PODs, MINOS has revealed a seasonal and a geographical gradient of harbour porpoise occurrence in the German Baltic EEZ. The numbers of POD-registered encounters with harbour porpoises show a general decrease from the island of Fehmarn in the West to the Pomeranian Bight in the East. In addition, seasonal variations are observed around Fehmarn and along the Kadet Channel and its adjacent waters – with increasing numbers during spring/summer and declining numbers in winter.

8.3.3 Usefulness of T-PODs

The approach of airborne surveys is reaching the limits of its abilities in areas with low porpoise densities, as well as in regions so small that they are overflown too fast. For such cases, T-PODs have proved to be the best way to observe porpoises, since they remain in their positions for weeks or months, listening to the porpoises' echolocation signals round-the-clock. They can detect the presence of harbour porpoises, and determine the spatio-temporal dynamics of their habitat use. But they are not suitable for estimating absolute stock size. Based on the experience gained during MINOS, we strongly recommend the combined use of both aerial and T-POD surveys. The snap-shot like and wide-area airborne surveys and the long-term deployment of the stationary T-PODs are complementary methods, which each contribute vital pieces to the overall results.

During the first years that EIAs were carried out for German offshore wind farms, most wind energy companies lost considerable numbers of T-PODs, which were simply torn off or otherwise vanished. This was a major additional cost item. So in 2003, the obligation to use T-PODs was suspended from the so-called standard investigation concept for offshore wind farm EIAs (BSH 2003). Since then, major developments have taken place, considering both hardware and data analysis. MINOS has e.g. improved the anchorage of the T-PODs, which has already significantly reduced the risk of loss. The improved T-POD technology is an indispensable tool for monitoring the impacts of building and operating offshore wind farms, as it allows surveying areas over long periods, and even under adverse weather conditions.

8.4   Habitat Loss

During installation of offshore windmills, the main impacts on both seals and harbour porpoises are the noise from ramming and similar building operations, increased ship traffic to and from the construction site, installation of cables, and greater turbidity of the water. These attendant circumstances of construction may also impair or deter fish, the seals' and porpoises' prey. The disturbances could result in avoiding the area mainly due to noise but as well as to lack of appropriate food. During operation of offshore wind farms the windmills emit low-frequency noise into the water, and maintenance traffic adds to this noise pollution. At present, only little is known about how these impacts affect the marine mammals.

The two biggest offshore wind farms, Horns Rev and Nysted/Rødsand, are quite dissimilar with regard to their hydrography and bathymetry and the density of harbour porpoises near them. Porpoises are much less abundant in the Baltic, where Rødsand is located. This must be kept in mind when looking at the results (Tougaard et al. 2004).

8.4.1 Harbour Porpoises

Harbour porpoises clearly avoided the Rødsand facility during building operations: Henriksen et al. (2004) report that there was a drastic decrease in porpoise echolocation activity from baseline to construction period, in both the impact and the control areas. The reduction was greater in the impact area, which the authors assume to be due not only to the lower number of porpoises in the area, but also to possible changes in habitat use during the construction period. Driving steel sheet piles with a barge-mounted vibrator immediately resulted in a very distinct and significant short-term decline of echolocation activity in both the construction and the reference areas (10 km away), which lasted for some hours after each application (Henriksen et al. 2003). The results imply a spatial impact gradient decreasing with distance from the construction zone. It is unlikely that the observed effects were caused solely by natural variations in porpoise density, or by some other anthropogenic impact (Henriksen et al. 2004).

The pre-construction studies of Horns Rev showed that harbour porpoises were abundant on the entire reef. Construction monitoring showed pronounced effects on the behaviour and abundance of porpoises over both the short (hours) and medium (whole construction period) terms (Tougaard et al. 2003b). Although mitigation measures were carried out beforehand, the strongest effects, again, occurred during the monopile ramming: Distinct short-term reductions of echolocation activity were caused which could be observed in the impact as well as in the control area (up to 15 km away). So the animals had evidently escaped from the building site over a large area. Acoustic activities returned to higher levels about three to four hours after the end of each ramming process, indicating that porpoises then entered the investigation area again. Moreover, behavioural changes could also be observed over large distances, with significantly fewer porpoises showing foraging behaviour (Tougaard et al. 2003b).

At present, it is very difficult to draw any further conclusion regarding the effects of the Horns Rev wind farm (Tougaard et al. 2005). There are contradictory findings: On the one hand, lower numbers of porpoises were visually observed inside the wind farm area during construction; on the other, there was a relative increase in porpoise echolocation activities inside the same area during the same period (except during ramming) (Tougaard et al. 2004). The increase could be explained by exploratory behaviour, or be due to natural influences. However, there is not enough data for any certain interpretation (Tougaard et al. 2005).

In 2003, the first year of operation, extensive service work became necessary at the Horn Rev wind farm due to technical problems. However, the resulting intensive ship traffic in the wind farm area seemed to have only little effects on porpoise abundance and acoustic activity, compared with the severe effects during construction in 2002 (Tougaard et al. 2004). It might thus be concluded that the main disturbing factor during construction consisted of the actual building activities, rather than the enhanced ship traffic.

8.4.2 Harbour Seals

The Rødsand seal sanctuary, 4 km away from the Nysted wind farm, is the most important resting site for harbour seals in the south-western Baltic Sea, and is also a transitory haul-out site for grey seals. The pre-construction study indicated that the wind farm area itself seems to be of less importance to harbour seals, and even lesser to grey seals (Dietz et al. 2003). There are no new data available to date. The conditions at Rødsand cannot be applied to German offshore planning sites, which are lying far away from seals' haul-out sites.

At Horns Reef, only a short baseline study of harbour seals was performed prior to construction (Tougaard et al. 2003a). It indicated that the entire Horns Reef appears to be an important corridor to the offshore foraging grounds. The present stage of extension of the wind farm seems to provide no barrier effect within this passage (Tougaard et al. 2003a). Current data indicate no negative influences on seals, such as large scale displacement during construction or operation. However, clear evidence is lacking because the spatial and temporal resolution of data to date is too limited (Tougaard et al. 2003a). Seven of eight seals traced crossed the wind farm area, but none seemed to spend any considerable time there.

8.5   Impairment of Hearing

As outlined above (see page 35) it is undisputed that marine mammals are extraordinarily dependent on their hearing system, e.g. for intra-specific communication, foraging, and orientation. They are thus particularly sensitive to noise emissions.

It is also beyond dispute that the installation of offshore windmills results in construction noise involving short-term activities of a few hours' duration, but these are repeated over a period of months at each wind farm site. Increased ship movement due to construction and maintenance, too, adds noise. Subsequently, the operation of an offshore wind farm brings low-frequency underwater noise with it. Compared to harbour porpoises, seals have good low-frequency hearing, and the males use it for intra-specific communication sounds (Richardson et al. 1995). Unfortunately, the question as to whether or to what extent disturbing noises affect the communication of marine mammals and impair or permanently damage their hearing is still largely unresolved.

With regard to single offshore windmills, Koschinski et al. (2003) exposed free-ranging harbour porpoises and seals to the simulated noise of a 2 MW wind turbine, and both species showed distinct reactions to this low-frequency sound. They were cautious and kept their distance from the source of the sound, but did not show any fear behaviour, as had been observed in former studies with deterring devices such as pingers. The porpoises in fact showed even more acoustic activity, which may be seen as an exploratory behaviour towards the new sound (Koschinski et al. 2003). Nevertheless, these reactions may be different with modified frequencies due to other turbine types or at stronger winds.

MINOS has succeeded in implementing the ABR method to evaluate possible noise impacts. The recording of Auditory Brainstem Responses (ABRs) may now be applied in further research. With more practical experience, it could become a future tool for ecological monitoring programmes. The measuring system can be used for in-air and underwater hearing, it may be applied to seals and porpoises in captivity as well as for seals in the wild. MINOS recorded a complete audiogram of a seal in the air, and a partial hearing curve of a porpoise underwater. The resulting thresholds, which, on the whole, are within the expected range, show the suitability of the chosen method for answering the questions posed.

The observations during the construction works of the two big Danish wind farms clearly indicated that harbour porpoises immediately responded to the pile driving noise by both deviant behaviour and escape from the building site and adjacent areas (Henriksen et al. 2004 and Tougaard et al. 2003b). The mitigation measures prior to ramming at Horns Rev included both a gradual ramp-up procedure and porpoise pingers and seal scarers. However, the benefit from using pingers and scarers is discussed controversially (Scheidat and Siebert 2003): They are on the one hand able to scare off the animals for their own good. However, as mentioned above, the noise emitted by those devices may indeed cause stress or harm to the mammals.

In general, even if an animal is apparently habituated to a disturbing noise, it does not necessarily mean that it has suffered no physical impairment. Such impacts could result in such effects as temporary or permanent threshold shifts. These effects as well as an increasing background noise could impair both intra-specific communication and general orientation – with a multitude of risks on population levels (Scheidat and Siebert 2003). The consequences of behavioural changes such as those observed during the ramming stages at Horns Rev could be manifold, and are in principle hardly assessable in the short or medium term.

8.6   From a Different Angle

However, offshore wind farms do not necessarily involve only disadvantages for marine mammals; rather, possible benefits due to expected reduction of fishing activities in offshore wind farm areas are also being discussed. These areas might provide a kind of refuge for benthic organisms, fish, and mammals, etc. The new hard substrates of the pylons could serve as artificial reefs, attracting new organisms – although displacing some of the original sandy bottom flora and fauna at the same time. If that led to a higher number of appropriate prey for marine mammals, positive effects might occur. Nevertheless, the above mentioned concerns regarding habituation effects must be kept in mind when debating possible benefits.

8.7   Conclusion

Due to the multitude of planned wind farms, the consequences on the flora and fauna cannot be ascertained by considering only a single wind farm. Instead, cumulative effects must be taken into account. A reasonable interval between the construction phases of separate wind farms could be a valuable measure to minimise impacts on the marine environment. Licences for German offshore wind farms therefore contain a condition which allows the licensing authority, the Federal Maritime and Hydrographic Agency (BSH), to coordinate the construction work. It is questionable whether there will be enough alternative and adequate low-noise areas for the animals if at least parts of their preferred habitats will be covered with windmills sometime. Another moot question is whether the animals will get used to the offshore wind farms or avoid them, presumably, that is a species specific matter. Even in theory it would be very difficult to assess the long-term impacts on reproduction and population status, and it cannot be done with our current state of knowledge. So we should have a close look at the future experience gained from the large existing and planned wind farms. Moreover, further studies on the sensory systems and on such environmental factors as oceanographic parameters are essential for a comprehensive picture of harbour seal and porpoise ecology, and for an understanding of how these animals react to changed conditions.

References

BioConsult SH and GFN (2002) Offshore-Windfarm Butendiek – Environmental Impact Assessment and Impact Assessment in relation to potential NATURA-2000 sites: On behalf of Offshore-Bürgerwindpark-Butendiek GmbH & Co. KG (OSB); English summary (June 2002). OSB, Husum, Germany. Available at: www.butendiek.de

BSH (Federal Maritime and Hydrographic Agency) (2003) Standards for Environmental Impact Assessments of Offshore Wind Turbines in the Marine Environment. Hamburg and Rostock, Germany. Available at: www.bsh.de

Diederichs A, Grünkorn T, Nehls G (2004) Einsatz von Klickdetektoren zur Erfassung von Schweinswalen im Seegebiet westlich von Sylt: 2nd Intermediate Report, 2004. Unpublished report commissioned by the Offshore-Bürgerwindpark-Butendiek GmbH & Co. KG. Bioconsult Schleswig-Holstein, Hockensbüll, Germany

Dietz R, Teilmann J, Henriksen OD, Laidre K (2003) Movements of seals from Rødsand seal sanctuary monitored by satellite telemetry: Relative importance of the Nysted Offshore Wind Farm area to the seals; NERI Technical Report No. 429. Ministry of the Environment. Copenhagen, Denmark. Available at: www2.dmu.dk/1_viden/2_Publikationer/3_fagrapporter/

Grünkorn T, Diederichs A, Nehls G (2004) Linientransektzählungen mit Flugzeug und Schiff westlich von Sylt: Fachgutachten Meeressäuger; Intermediate report covering May through Dezember 2003. Unpublished report commissioned by the Offshore-Bürgerwindpark-Butendiek GmbH & Co. KG. Bioconsult Schleswig-Holstein, Hockensbüll, Germany

Henriksen OD, Teilmann J, Carstensen J (2003) Effects of the Nysted Offshore Wind Farm construction on harbour porpoises: The 2002 annual status report for the acoustic T-POD monitoring programme; Technical report. NERI, Roskilde, Denmark. Available at: www.nystedhavmoellepark.dk/upload/pdf/NERI03_eohp_TPOD.PDF

Henriksen OD, Carstensen J, Tougaard J, Teilmann J (2004) Effects of the Nysted Offshore Wind Farm construction on harbour porpoises: Annual status report for the acoustic T-POD monitoring programme during 2003; Technical Report to Energi E2 A/S. NERI, Roskilde, Denmark

Koschinski S, Culik BM, Henriksen OD, Tregenza N, Ellis G, Jansen C, Kathe G (2003) Behavioural reactions of free-ranging porpoises and seals to the noise of a simulated 2 MW windpower generator. Mar. Ecol. Prog. Ser. 265, pp. 263-273

Richardson WJ, Greene CR, Malme CI, Thomson DH, Moore S, Würsig B (1995) Marine mammals and noise. Academic Press, London: 576p.

Scheidat M, Siebert U (2003) Aktueller Wissensstand zur Bewertung von anthropogenen Einflüsse auf Schweinswale in der deutschen Nordsee. Seevögel, Zeitschrift Jordsand, Vol. 24, No. 3, pp. 50-60

Siebert U, Kock KH, Scheidat M (2003) Erfassung der Dichte und Verteilungsmuster von Schweinswalen (Phocoena phocoena) in der deutschen Nord- und Ostsee: Teilprojekt 2 – Schweinswal-Dichten. In: National Park Regional Office for the Schleswig-Holstein Wadden Sea (ed). First Interim Project Report MINOS. Funding code at German Federal Environment Ministry FKZ 0327520. Available at: www.minos-info.org

Siebert U, Kock KH, Scheidat M (2004) Erfassung der Dichte und Verteilungsmuster von Schweinswalen (Phocoena phocoena) in der deutschen Nord- und Ostsee: Teilprojekt 2 - Schweinswal-Dichten. In: National Park Regional Office for the Schleswig-Holstein Wadden Sea (ed). Final Project Report MINOS. Funding code at German Federal Environment Ministry FKZ 0327520. Available at: www.minos-info.org

Tougaard J, Ebbesen I, Tougaard S, Jensen T, Teilmann J (2003a) Satellite tracking of harbour seals on Horns Reef. Technical report to Techwise A/S, Biological Papers from the Fisheries and Maritime Museum, Esbjerg. No.3

Tougaard J, Carstensen J, Henriksen OD, Skov H, Teilmann J (2003b) Short-term effects of the construction of wind turbines on harbour porpoises at Horns Reef. Technical Report to TechWise A/S. HME/362-02662, Hedeselskabet, Roskilde/Denmark

Tougaard J, Carstensen J, Henriksen OD, Teilmann J, Rye Hansen J (2004) Harbour Porpoises on Horns Reef: Effects of the Horns Reef Wind Farm; Annual Status Report 2003 (final version June 2004). NERI Technical Report to Elsam Engineering A/S. Available at www.hornsrev.dk

Tougaard J, Carstensen J, Wisz MS, Teilmann J, Ilsted Bech N, Skov H, Henriksen OD (2005) Harbour Porpoises on Horns Reef: Effects of the Horns Reef Wind Farm; Annual Status Report 2004, July 2005. NERI Technical Report to Elsam Engineering A/S. Available at: www.hornsrev.dk

Research on
Bird Migration

Background

Annually, more than ten million birds cross the North and Baltic Seas, both during the spring and the autumn migrations, on their way between their breeding and their winter habitats. Since the North and Baltic Seas are central not only to the European, but also to global bird migration patterns, both are of outstanding international importance.

Observations to date have indicated that the bird migration, particularly over the North Sea, proceeds largely in the form of broad-front migration, i.e. there are no particular corridors preferred by the migratory birds. Only under special weather conditions do the guideline effects of coasts or rivers come into play. In addition to seasonal fluctuations (a few days with extremely high intensity), the migratory intensity is also subject to very strong diurnal fluctuations. Depending on species and weather conditions, migratory birds fly at very different heights, ranging from just over the water surface to several thousand metres' altitude.

The construction of offshore wind parks will introduce vertical structures to a hitherto largely vacant area located at the migratory altitude of many bird species. The main danger area is the range between 20 and 200 m height. In view of the dimensions of the offshore wind parks being proposed for the North and Baltic Seas, endangerment to bird migration basically cannot be precluded.

Potentially two particular correlations of effects exist which could constitute endangerment to bird migration:

- the danger of collision with the turbines (bird strike), and
- the barrier effect, with forced avoidance and circumnavigation of the parks by the birds, resulting in an increased consumption of energy reserves.

While a large number of investigations into the effects of wind turbines on birds on land have been carried out, there have to date been only a few experience reports regarding the offshore area. The monitoring results of the two Danish offshore wind parks, which are comparable in size with the wind parks proposed in German waters, have provided initial indications. By contrast to the temporary impairments during the construction phase, the effects on bird migration during the operational phase are permanent.

The endangerment level involving bird migration primarily depends on the number of individuals concerned. This in turn depends not only on flight altitude and time of day, but also on species-specific avoidance and circumnavigation behaviour, specific weather conditions (fog, wind direction, and wind speed), and also the size and extent of the wind park. However, no certain quantification of losses to be expected is possible to date. In future, we will be better able to forecast the number of bird collisions; moreover, in addition to a more precise investigation into numbers and species compositions, an investigation of distribution in space and time will also be necessary. An assessment of any possible effects on populations will require the development of population models, including such species-specific demographic parameters as stock size, distribution, reproduction rate and mortality rate.

The research into the bird migration over the North and Baltic Seas will therefore deal with issues of the progression of intensity of bird migration, altitude distribution depending on the weather, and a further-reaching analysis of migration directions.

9 Bird Migration and Offshore Wind Turbines

Ommo Hüppop, Jochen Dierschke, Klaus-Michael Exo, Elvira Fredrich,
Reinhold Hill

9.1 Introduction

Worldwide, Germany became the leading country in the use of wind energy. Since most suitable land-based sites for wind turbines are now occupied, and winds at sea are generally stronger and more stable, ambitious plans for offshore sites have been broached. There are now applications for 33 sites within the German Exclusive Economic Zone (EEZ) in the North and Baltic Seas, some calling for several hundred turbines. As several hundred million birds cross the North and Baltic Seas at least twice each year, the Marine Facilities Ordinance stipulates that licenses will not be issued to facilities which jeopardise bird migration.

Birds are potentially endangered by offshore wind farms through collisions, barrier effects and habitat loss. To judge these potential risks, the occurrence of birds in space and time as well as details on their behaviour in general and their behaviour at wind farms need to be determined. The effects of construction and maintenance work must also be considered.

Since 2003, we have investigated year-round bird migration over the North Sea with regard to offshore wind farms. The results of measurements with radar, thermal imaging, and visual and acoustic observations have been compiled now. This chapter is a shortened and modified version of an article that appeared in early 2006 in "Ibis", the International Journal of Avian Science published by the British Ornithologists' Union (Hüppop et al. 2006).

Of the 33 projected wind farm sites in the German EEZ, 27 are in the North Sea and six in the Baltic Sea; some involve several hundred individual turbines. Twelve pilot projects with between twelve and 80 turbines each have now been approved (ten in the North Sea, two in the Baltic Sea). Two others in the Baltic Sea were rejected because of large concentrations of resting birds in their respective areas. Shoreward of the EEZ, i.e. in coastal waters inside the 12-nm zone, permits have been granted for additional wind farms (for details see, for example, http://www.offshore-wind.com). Even a far more modest development would make the construction of wind farms the greatest human impact in the North and Baltic

Seas next to fisheries (Merck and von Nordheim 2000). However, construction has not yet started at any licensed site in Germany.

Such interests as mining rights, shipping routes, the navy, commercial fishing and nature conservation, as well as those of submarine cable and pipeline operators, and safety issues must all be considered before a farm is licensed. One of the reasons for rejection explicitly stated in the Marine Facilities Ordinance is "jeopardising bird migration" (Dahlke 2002). Approval may not, however, be withheld without reasons. But what does "jeopardising bird migration" mean in terms of, for example, numbers of collision victims or effects and impacts on bird populations?

There exists a comprehensive literature on bird migration over the North Sea from the end of the 19th century onwards (e.g. Gätke 1891). It includes both visual observations (see already Drost 1928) and extensive technical approaches such as surveillance radar studies by Lack and others (reviewed by Eastwood 1967, for the Helgoland Bight of the North Sea, e.g. Jellmann and Vauk 1978) or satellite telemetry (Green 2003). Nevertheless, with respect to questions regarding environmental effects and impacts connected with the construction of offshore wind turbines, severe gaps in our knowledge have become evident:

1. How many migrants of which species cross the German Bight at which times?
2. What is the proportion of birds flying in altitudes up to 200 m (the approximate height of the projected wind turbines)?
3. How are migration intensity and flight altitude influenced by weather – namely by wind, precipitation and visibility?
4. How many birds are involved in reverse migration?
5. How do migrants react to anthropogenic offshore obstacles?
6. Are birds attracted by the illumination of these structures?
7. How many birds will collide?
8. Can days of high collision risk be predicted?
9. How can collisions be mitigated?
10. Which impacts on populations can we expect?

Since roughly $^2/_3$ of all bird species migrate during darkness, when the collision risk with wind turbines is expected to be higher than during daylight, special techniques for studying this "invisible migration" have to be applied. Most data and analyses presented here are derived from the project BeoFINO, the primary objectives of which were to collect data on issues (1) through (7), based on measurements of bird migration over the German Bight, using a variety of techniques, including radar, thermal imaging, collection of collision victims, and visual and acoustic observations.

This is the first project to cover migration year-round with such a variety of complementary methods.

9.2   Methods and Data

The BeoFINO project was centred at the FINO 1 platform about 45 km north of the island of Borkum (54° 01' N, 06° 35' E; http://www.fino-offshore.com), and, starting in October 2003, conducted remote observations, including those of "invisible" bird migration, via two ship radars, a thermal imaging camera, a video camera and a directional microphone. To allow spatial comparisons, it also included data from human observers on islands.

In the North Sea, two periods of intensive migration (spring, autumn) recognisably alternate with two periods of less intensive migration (summer, winter). For our purposes, these seasons were defined as follows: spring: 1 March - 31 May; summer: 1 June - 31 July; autumn: 1 August - 15 November; and winter: 16 November - 29 February. A meaningful subdivision of the day, taking into account fluctuating day lengths, proved more of a problem. We settled on a compromise between civil dawn/dusk, when the sun is 6° below the horizon, and nautical dawn/dusk, when it is 12° below the horizon: we used sun declinations of −9° and +9°, respectively, to subdivide a day, yielding four periods referred to herein as morning, daytime, evening, and night. Nights are assigned to the date on which they begin.

9.2.1   Sea-Watching and Passerine Passage Counts

Standardised systematic recordings of "visible" bird migration (alternating "sea-watching" recordings of waterbirds over sea, and "passerine passage counts" recordings of passerines, pigeons, owls, swifts and woodpeckers) were carried out by observers on the offshore island of Helgoland and simultaneously on the coastal islands of Sylt and Wangerooge in 2003 and 2004 (for locations see Hüppop et al. 2006, for methods see Hüppop et al. 2004 and Dierschke et al. 2005). On Helgoland, sea-watching was carried out on 233 days in all seasons, and passerine passage counts on 90 days from July 2003 to December 2004. On Sylt, counts were conducted on 156 (sea-watching) and 98 (passerine passage count) days in autumn 2003, spring 2004 and autumn 2004. On Wangerooge, 90 sea-watching and 58 passerine passage count days were conducted in spring and autumn 2004.

9.2.2 Ship Radar

One radar system with a vertically rotating antenna recorded bird migration intensity and altitude at the FINO 1 platform (Fig. 1) from the middle of October 2003 onwards. A second radar, rotating horizontally, operated on FINO 1 starting 30 October 2003, to record flight directions; it scanned the sea north of the platform from 225° to 135°. Another vertical radar was in operation at the airfield on the island of Sylt, scanning west-southwest and north-northeast from 8 June 2004 to 6 November 2004, to compare coastal and offshore migration density and flight altitudes; for these purposes, no horizontal radar was necessary on Sylt.

Fig. 1. A ship radar on the research platform FINO 1 was tilted by 90° to record flight intensities and altitudes (left). A second one operates in its normal position to study flight directions and later, during construction and operation of the nearby pilot wind farm, possible effects on flight behaviour of migrating birds (right). Photographs by R. Hill

This article is mainly based on data collected from 1 October 2003 to 15 November 2004 (vertical radar), and from 1 March 2004 to 31 May 2004 (horizontal radar). Since a web camera (http://www.fino-offshore.de) occasionally showed large numbers of resting gulls on FINO 1, we assume that bird echoes outside the "night" period are partly attributable to foraging gulls. By contrast, the number of traces recorded at night is in agreement with the phenology described by Hüppop and Hüppop (2004) for passerines on Helgoland, so that the vast majority of signals recorded in this period are undoubtedly attributable to migration. Therefore, some analyses were exclusively based on echoes recorded at "night", so as to minimise influence of other flight activities.

To record bird echoes, screenshots were taken of the radar image at five-minute intervals using IrfanView software (www.irfanview.com) and subsequently digitised by hand. Despite occasional system crashes, namely in the beginning, with concomitant data losses, the vertical radar was running for 67 % of the investigation period on FINO 1, and generated a total of almost 80,000 images. 69 % of the investigation period was covered on Sylt. All images that were more than 20 % obscured by rain reflections (7.8 % of the images on FINO 1 and 7.3 % on Sylt) were discarded. Short breaks in radar recording also occurred, in which case only those periods of the day were included in the analysis which provided at least 50 % of the total number of images that were theoretically possible. Altogether, 62 % of the total time was covered by images suitable for analysis, both on FINO 1 and on Sylt. Of the 412 "nights" in the 13.5-month recording period, 226 were entered into the analysis. Most of the missing nights were due to lengthy recording gaps caused by radar system breakdowns, which affected some annual segments more than others. Nonetheless, the radar measurements provide an unique almost continuous account of offshore bird migration throughout the annual cycle.

Birds flying parallel to a radar sweep left tracks of dots (Fig. 2), while those flying perpendicular to it left a single dot. This information was manually digitised from the screenshots. Migration intensities and flight altitudes could be derived from all echoes, while track inclinations and rough flight directions were derived only from tracks at least 35 m long (equal to 10 pixels in an image, or roughly five times the radar's spatial resolution).

The number of echoes recorded had to be corrected for the change in detectability with distance from the radar antenna, which was done via the program Distance 4.1 (www.ruwpa.st-and.ac.uk/distance/), based on the assumption that birds flying over the sea are horizontally equally distributed at such a small scale that e.g. knowing radar cross-sections of the birds is unnecessary (Hüppop et al. 2002, 2006). All values from altitudes between 50 and 150 m and at a distance of over 400 m from the platform were included, so as to reduce data of gulls resident on the platform. Only a very few bird echoes were recorded from distances over 1,500 m, and these may have been over-adjusted in the correction. Hence, our calculations include only values within a radius of 1,500 m (Hüppop et al. 2004).

Fig. 2. Part of the display of the vertical radar (October 2005). The yellow bottom line results from radar reflections from the sea surface, the "y-axis" denotes altitude. The distance between the white rings is ¼ nautical mile (= 463 m). Several tracks of nocturnally migrating birds heading southwest in different altitudes are visible. Each dot represents an echo from one radar sweep (24 sweeps per minute), the yellow ones being the most recent echoes

The radar images of the horizontally operating radar were largely obscured by wave and rain reflections, so that only 4.1 % of all images were in fact digitised, which was not sufficient to yield usable results. Improvements in antenna design may lead to better experiences in future.

9.2.3 Thermal Imaging, Video Camera and Microphone

Thermal imaging cameras make the long-wave infrared radiation of birds and other heat radiators visible (e.g. Desholm and Kahlert 2005). This enabled us to detect movements near the platform around the clock without additional light, and thus record intensity, flock size and flight behaviour under various environmental conditions (Fig. 3). We used a Zeiss Optronics Opus M camera (for details see Hüppop et al. 2006). The field of view of its 75-mm lens was 12° x 9° to the north, with an elevation angle of 60°

to the open sky. We developed new software, IRMA (Infrared Registration of Migrating Aves) to automatically detect flying birds or bats within the real-time images of our thermal imaging camera.

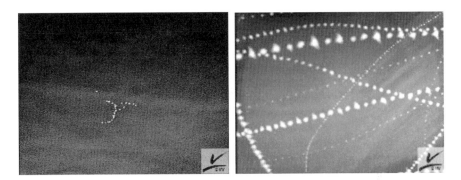

Fig. 3. Thermal images of nocturnally migrating ducks / geese (single image, left) and of small birds that fly disoriented around the platform FINO 1 in a night of bad visibility (more than 3.000 single images stacked over 5 minutes using peak storage, right)

During daylight, a video camera (Panasonic AWE600E) with a motor zoom lens on a pan-and-tilt head was used as our "sea-watcher" on the unmanned platform. Unfortunately, we were not able to observe birds remotely because of insufficient resolution, combined with a limited field of view and Internet bandwidth problems. Bird calls close to the platform were detected and recorded automatically by a directional microphone (Sennheiser ME67), with the specially developed software AROMA (Acoustic Recording of Migrating Aves, Hüppop in prep.).

9.2.4   Collision Victims

Investigations of collisions with man-made offshore structures had to be confined to the FINO 1 platform (due to the lack of offshore wind farms in German waters). All bird carcasses found during 44 visits to the platform from October 2003 to December 2004 were documented. Most of them were taken to the laboratory, where an attempt was made to establish the cause of death by examining each individual for external injuries. A few birds were additionally X-rayed.

9.3   Results

9.3.1   Species Composition

On Helgoland, more than 425 species have been recorded, which shows the great number of species which migrate across the North Sea. However, the proportions of any one species can be estimated only very crudely, as many birds are mainly or exclusively nocturnal migrants, their "noise levels" depending on species, visibility conditions, etc. (Alerstam 1990).

The systematic visual daytime observations in 2003-2004 ascertained a total of 217 species (192 on Sylt, 174 on Wangerooge and 167 on Helgoland). At all sites, waterfowl, including great cormorant *Phalacrocorax carbo*, gulls and terns were the dominant groups recorded by sea-watching (Hüppop et al. 2006), with great cormorant, greylag goose *Anser anser*, pink-footed goose *Anser brachyrhynchus*, barnacle goose *Branta leucopsis*, brent goose *Branta bernicla*, common eider *Somateria mollissima*, common scoter *Melanitta nigra*, common gull *Larus canus*, black-headed gull *Larus ridibundus*, lesser black-backed gull *Larus fuscus*, sandwich tern *Sterna sandvicensis* and common tern *Sterna hirundo* being numerically the most important species. Passerine passage counts showed a similar species spectrum at all three locations (Hüppop et al. 2006). Common wood pigeon *Columba palumbus*, meadow pipit *Anthus pratensis*, white wagtail *Motacilla alba*, fieldfare *Turdus pilaris*, redwing *Turdus iliacus*, song thrush *Turdus philomelos*, Eurasian jackdaw *Corvus monedula*, brambling *Fringilla montifringilla*, chaffinch *Fringilla coelebs* and common linnet *Carduelis cannabina* were the most numerous species.

At the FINO 1 platform, 70 different species were verified by the automatic flight call recording. Over 70 % of the registered flight calls (n = 19,776) were thrushes, chiefly redwings, blackbirds *Turdus merula*, fieldfares and song thrushes; some 10 % were waders, primarily common redshanks *Tringa totanus*, red knots *Calidris canutus*, Eurasian golden plovers *Pluvialis apricaria*, common sandpipers *Actitis hypoleucos* and greenshanks *Tringa nebularia*. Other frequently registrated species were sky lark *Alauda arvensis*, meadow pipit, goldcrest *Regulus regulus*, European robin *Erithacus rubecula*, common starling *Sturnus vulgaris* and snow bunting *Plectrophenax nivalis*, although most of the calls made by the last four species in all likelihood involved birds resting on the platform.

A total of 13,037 birds were recorded by the thermal imaging camera at night. There were six mass migration events with more than 500 individuals each. The density of birds near the platform was so high on these nights that it was impossible to distinguish between flocks and single birds, so that 10,340 birds in these nights were not included in flock size calculations. Of the remaining 2,697 individual records of night migrants, 763 (28.3 %) could be roughly identified, of which 52.2 % were passerines. Over 94 % of these were single birds (80 %).

Of the 442 carcasses found on FINO 1, only six were non-passerines (one dunlin *Calidris alpina*, four large gulls, and one feral pigeon *Columba livia*). Most of the birds involved were thrushes (87.3 %), common starlings (4.8 %) and sky larks (1.6 %).

9.3.2 Seasonal Migration Intensities

All methods used confirmed year-round bird migration with a lot of both, seasonal and day-to-day variation. Sea-watching at Helgoland yielded migration peaks in spring and autumn (Fig. 4). In addition, there were some noticeable movements during summer and winter, which were not recorded on Wangerooge and Sylt because of the reduced observation times. The lowest migration intensity was found in mid-June. Most ducks were observed from September to April, terns dominated in late spring and early autumn, while gulls were most numerous during winter.

In contrast, passerine migration was much less spread over the year. It was strongest in spring on Wangerooge and in autumn on Sylt, which emphasizes coastal effects. Diurnal passerine migration intensity on Helgoland was generally lower compared to the two coastal islands (Fig. 5). Regarding the main groups, differences in the time of migration were less pronounced than between the groups recorded by sea-watching (Hüppop et al. 2006)

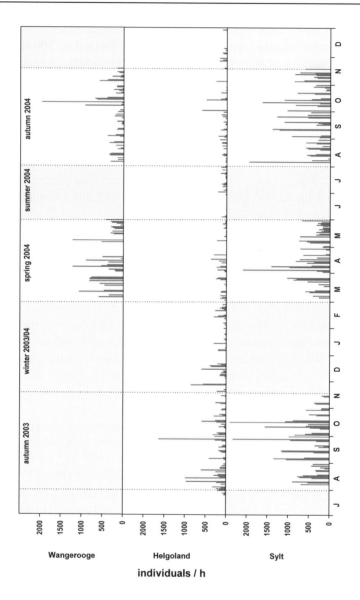

Fig. 4. Variation in daily migration intensity recorded by sea-watching on Wange-rooge (n = 85.538 individuals), Helgoland (n = 87.098) and Sylt (n = 238.765) from July 2003 to December 2004. Note that there was no recording on Wange-rooge and Sylt in summer and winter (indicated by grey fields) and that observations were not carried out every day (for details and time schedule see Hüppop et al. 2006)

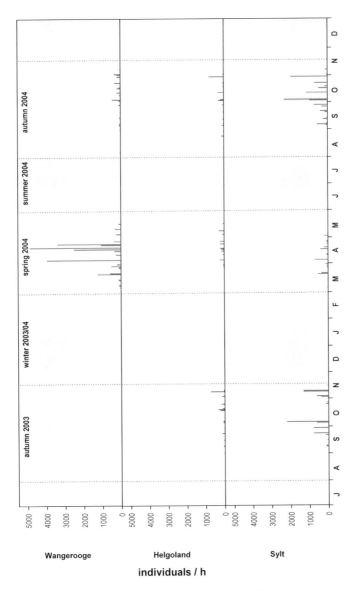

Fig. 5. Variation in daily migration intensity recorded by passerine passage counts on Wangerooge (n = 70.302 individuals), Helgoland (n = 21.908) and Sylt (n = 67.670) from August 2003 to November 2004. Note that there was no recording on Wangerooge and Sylt in summer and winter (indicated by grey fields) and that observations were not carried out every day (for details and time schedule see Hüppop et al. 2006)

Nocturnal migration registered by vertical radar also showed large seasonal and day-to-day differences of intensity (Fig. 6). Periods of higher migration density were October/November 2003, mid-March through early May 2004, and November 2004. Hardly any bird migration was detected from December 2003 to February 2004, or in June and August 2004. Long radar breakdowns occurred in May/June and especially October 2004. In autumn 2003, part of the bird migration came unusually late (end of November), with intensive diurnal migration of mainly greylag geese, Eurasian wigeons *Anas penelope*, red-breasted mergansers *Mergus serrator*, black-headed gulls and blackbirds recorded on Helgoland.

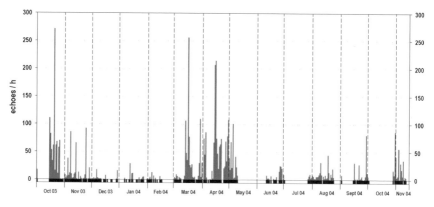

Fig. 6. Migration intensities based on radar echoes corrected for distance effects from October 2003 to November 2004 at research platform FINO 1 (n = 10,972 echoes). Black bars under the histogram indicate days with sufficient numbers of radar measurements

It is noticeable that in "visible" spring migration, densities are higher at winds from the east, south and south-west, but not at winds from the north-west, north and north-east. In autumn, more intensive migration occurs at winds from the north-east and east, and less at winds from the south-west and north. In summary, there is a higher migration intensity during tailwinds, both in spring and autumn, although too much tailwind leads to a reverse effect (Fig. 7).

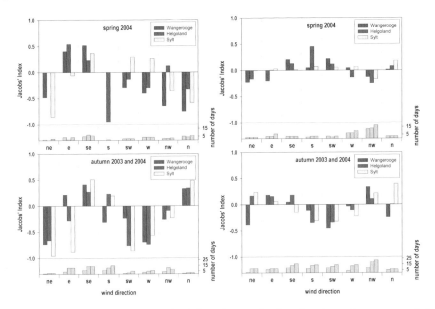

Fig. 7. Relationship between wind direction and migration intensity as recorded by sea-watching (left) and by passerine passage counts (right) at the islands of Helgoland (offshore), Wangerooge and Sylt (both coastal). Positive values of the Jacobs' Index (Jacobs 1974) denote a preference, negative ones an avoidance of the respective wind-direction. Note different numbers of observation days at the three sites, namely in autumn

9.3.3   Daily Variation of Migration Intensities

In all migration periods, bird movements varied substantially from day to day and were concentrated in a few days and nights (Fig. 4 to 6). In particular, pink-footed geese were recorded in such high concentrations by sea-watching that over 75 % of individuals migrated on only 2 - 5 % of observation days. Considering all species observed by sea-watching, 75 % of individuals were observed on only 39 - 48 % of days, and 50 % on only 17 - 25 % of days (Fig. 4). The visible migration intensity of passerines showed an even higher variability and concentration. Half of all passerine passage was recorded on only 6 - 21 % of observation days, and 75 % on 17 - 33 % of days, depending on location and season (Fig. 5). The visual findings were confirmed by radar. In spring, more than half of all echoes were registered in only eight nights (25 % of echoes in three nights, 75 % in 18 nights). In autumn 2003 (15 October to 15 November), 50 % of the echoes were recorded in five of 31 observation nights, and in autumn 2004 (1 August to 15 November) in six of 61 nights (Fig. 6).

The relationship between migration intensity and wind is complex (see example in Fig. 8): Periods of strong headwinds normally keep the migrants on the ground. A sudden advancement in wind condition may then lead to a "discharge" of migrants often in one single night, before migration intensity drops again (e.g. mid March 2004). During longer periods of unfavourable conditions, even small changes in flight conditions may cause such "eruptions" (e.g. mid April 2004).

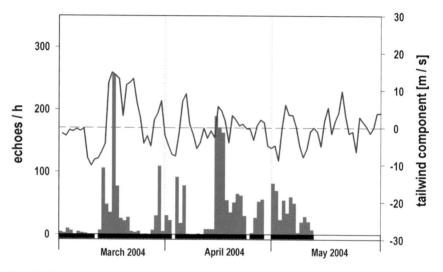

Fig. 8. Nocturnal migration intensity at FINO 1 recorded by vertical radar in relation to the tailwind component (n = 4,907 echoes, black bars under histogram denote days with sufficient sample size). The "tailwind-component" is a measure for wind force and direction with respect to the flight direction (Fransson 1998). Highest values mean strongest tailwinds, lowest values strongest headwinds

There were major differences between the migration densities registered by vertical radar on FINO 1 and on Sylt. In autumn 2004, data were collected simultaneously at both sites in 41 nights. Bird migration on FINO 1 was greater in only six nights. In one night, no activity was registered at either site, while in the remaining 34 nights, the activity on Sylt was distinctly greater. In summary, nocturnal migration in autumn 2004 was markedly greater on Sylt than at FINO 1 (Wilcoxon test, z = –4.342, p < 0.001). The influence of wind direction on migration density was very similar at both sites. Differences occurred at east winds (preference on FINO 1, avoidance on Sylt), and at north and west winds (avoidance on FINO 1, preference on Sylt).

There was good agreement between call and thermal imaging records (for all nights with at least one bird call/infra-red record: $r_s = 0.385$, $p < 0.02$, $n = 37$ nights), but hardly any with intensities recorded by vertical radar. High numbers of birds near the platform obviously coincide with drizzle or mist, but the database is still too small for a more detailed analysis.

9.3.4   Daytime Variation of Migration Intensities

"Visible" migration was strongest at all locations and in all seasons during the first three hours after sunrise, with the third hour often showing clearly lower migration intensities (Hüppop et al. 2006). At noon and in the three hours before sunset, migration intensity apparently dropped. This daily rhythm was weakest in spring on Helgoland and Sylt, and in summer and winter on Helgoland.

The adjusted bird echoes derived from the vertical radar revealed that the majority of "invisible" bird movements were recorded at "night", with almost always lower relative migration densities in the morning and evening (Hüppop et al. 2006).

Most birds picked up by the call recording system and by the thermal imaging camera were recorded at night (75.6 and 77 %, respectively), not including resting gulls and sandwich terns.

9.3.5   Migration Altitude

On average, almost half of the echoes up to 1,500 m came from the lowest 200 m, and thus from birds that flew directly within the activity radius of future offshore wind turbines (Fig. 9). Above this level, the number of echoes dropped in most cases. Whatever the time of day or season, the highest percentage was almost always registered in the lowest 100 m, particularly in daytime and to a lesser extent in the morning and evening periods. Since many echoes at these times presumably relate to gulls (see above), only birds migrating at night are considered in the following analysis. At night, most birds also migrate at altitudes below 200 m, ranging from 20 % of all echoes in summer 2004 to 64 % in winter 2003-2004. Most of these winter echoes were due to the late onset of migration in the second half of November, so that altitude distribution is closer to that of autumn. In spring, the concentration in the lower strata is less evident (Hüppop et al. 2006). Due to a shortage of data no interpretation can be made for the summer.

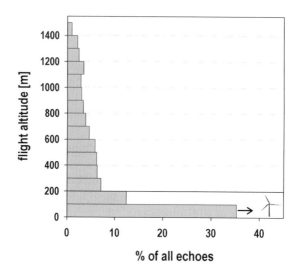

Fig. 9. Average distribution of flight altitudes up to 1,500 m (percentage of corrected echoes per altitude stratum, n = 23,814 uncorrected echoes) as measured with vertical radar at the FINO 1 research platform. Echoes of low flying birds cannot be separated from reflections of the sea surface. Thus, the arrow emphasizes that the proportion of echoes in the stratum from 0 to 100 m is presumably greater. The line at 200 m indicates the approximate maximum height of the future offshore wind energy plants

In rainy nights, the percentage of birds migrating below 200 m was significantly higher than in nights without rain (33 vs. 25 % of all echoes up to 1500 m, G-test, P < 0.01).

In both spring and autumn, tailwinds and light headwinds were associated with higher flight altitudes. With tailwinds above certain strength, flight altitudes tended to drop off again. In autumn, the five nights of heaviest migration were associated with easterly winds, in spring with southerly winds. However, the distribution of flight altitudes among the various nights could hardly be more disparate (own unpubl. data) and will require additional data and analysis.

9.3.6 Reverse Migration

Birds migrating roughly from the south-west to the north-east (main direction in the German Bight during spring migration) or vice versa (main direction in the German Bight during autumn migration) should generate obvious tracks at the vertical radar at FINO 1. Rough estimates (i.e. north to east vs. south to west) of flight direction were made using vertical radar, and, for the analysis, a track length of at least 35 m (see above). This permits determination of the proportion of reverse migration. At night in spring, 92.5 % of the birds with pronounced tracks migrated north- to eastwards, while 7.5 % migrated in the opposite direction. In autumn, 70.1 % headed west to south and 29.9 % in reverse direction. The "right" direction predominates only at night and in the morning in spring and autumn (Hüppop et al. 2006). At other times of day and in the seasons of lower migration density, flight direction patterns were clearly dominated by local birds (e.g. gulls residing on or near the platform).

9.3.7   Spatial Distribution

The visual observations showed a clear concentration of waterfowl near the shore (Fig. 4), with only few species showing equal or larger numbers in the offshore area (e.g. red-throated diver *Gavia stellata*, pink-footed goose). In addition, passerines concentrated more strongly at the coast than offshore, which confirms prior investigations (Dierschke 2001). Migration further concentrated at the "departure coast", i.e., in spring on Wangerooge and in autumn on Sylt. At the appropriate migration times, there may be impressive concentrations of passerines inshore, while offshore migration is barely noticeable. Birds seem generally to cross the German Bight at offshore winds. Vertical radar measurements on Sylt and FINO 1 confirmed a higher intensity of "invisible" migration near-shore. Very recent analyses also indicate a considerably higher migration intensity at Helgoland than at FINO 1 (own unpubl. data).

9.3.8   Collisions

A total of 442 birds of 21 species were found dead at FINO 1 between October 2003 and December 2004. Nearly all were in good physical condition, which excludes starvation as a cause of death; 245 individuals (76.1 %) had visible injuries, most commonly bleedings at the bill (41.3 %), contusions on the skull, and broken legs (16.8 %). Possibly, some birds died of exhaustion caused by flying around the platform (Hope Jones 1980). Over 50 % of strikes occurred in just two nights (1 October 2003 and 29 October 2004), involving a total of 86 and 196 birds, respectively. Both nights had periods of very poor visibility, with mist or drizzle, and presumably an increased attraction by the illuminated research platform. In the second of these nights the thermal imaging camera revealed that many birds flew around the illuminated platform, obviously disoriented (in the first night the camera was not yet in operation).

Fig. 10. Redwings and a few song thrushes found dead at the research platform FINO 1 in October 2004. Feathers at the rail clearly indicate collisions

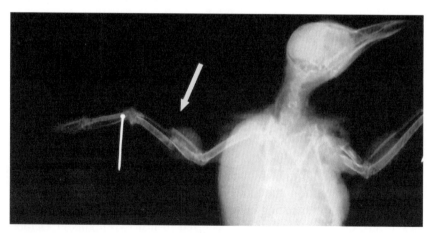

Fig. 11. Radiograph of a redwing found dead at the research platform FINO 1. Both, ulna and radius are fractured (arrow)

9.4   Discussion

9.4.1   Advantages and Disadvantages of Methods

Visual observation has traditionally been the only method which has permitted conclusions to be drawn on migration intensity by species. However, sea-watching covers only the lowest 100 m, and passerine passage counts only the lowest 200 m, since few birds can be detected at higher altitudes (Dierschke et al. 2005), and only in daytime at fair visibility.

Radar systems provide data independent of daytime and weather conditions, although the quality is affected by rain, and deteriorates with increasing distance and decreasing bird size (e.g. Sutter 1957, Eastwood 1967, Bruderer 1997a, 1997b). Rotating vertically, an ordinary ship radar can deliver a continuous record of both flight altitude and migration density (Cooper et al. 1991, Hüppop et al. 2004). But in contrast to tracking radar (Bruderer and Jacquat 1972), it is not yet adequate for species identification. Nevertheless, first experiments with a fixed antenna rendered promising results.

Low-flying birds can be hidden in surface or sea clutter. Furthermore, owing to the fairly limited range of a ship radar, it is necessary to know how many birds are missed at higher altitudes. Bruderer and Liechti (2004) showed that 50 % of all migrants over Switzerland and southern Germany could be detected at an altitude between 0 m and 250 or 750 m (depending on season and time of day; 90 % at 0 to 600 - 2000 m). Hence we believe that the birds missed, i.e. those flying above 1500 m, are a negligible number. Although horizontal ship radar can provide data on flight direction, such as avoidance flights at wind farms, an uninterrupted recording of migration at sea is only possible in the absence of waves (Hüppop et al. 2004).

Radar investigations specifically targeting offshore bird migration have also been incorporated into Environmental Impact Assessments (e.g. Gruber and Nehls 2003, Walter and Todeskino 2005), but are limited to a few nights per migration season.

The use of civilian or military surveillance radar tremendously improved the understanding of large-scale spatial distribution of bird migration (e.g. Eastwood 1967, Bruderer 1997a, 1997b). However, like weather radar, it is normally not installed for bird studies so that generally, only by-product data are available. In Germany, data from the Air Force's large aircraft surveillance radars are used for bird-strike warnings (Friebe 1998), but their use to quantify bird migration is limited by security considerations. The accuracy of altitude and time of migrating birds is limited by angle precision and spatial resolution, and, at greater distances, by the curvature of the earth's surface (e.g. Bruderer et al. 1995a). Still, with certain assumptions, spatial comparisons of migratory intensity and direction which would otherwise be impossible, can be made.

Although the species that cross the German Bight are well known, it is still a technical challenge to identify altitude distribution and migration intensity by species with remote techniques. Thermal imaging and video cameras have shorter range, but provide information on species and flock size. They are the only available method for recording birds flying disoriented around the platform at nights with poor visibility. They can also be used to record collisions (Desholm and Kahlert 2005), but the resolution of affordable models is too low, e.g. to detect a colliding passerine at 100 m distance. Collision numbers are therefore presumably higher than assumed. Carcass collections are unreliable at sea, as a large proportion of corpses fall into the water or are scavenged by gulls. The quantification of collision victims at sea therefore remains a technical challenge.

9.4.2   Migration Intensity, Altitude and Direction

Presupposing that the mass of birds migrate along a northeast–southwest axis over the German Bight, the autumn weather conditions in Scandinavia and the spring conditions in the Low Countries and in southern England have the largest influence on bird migration. Namely the wind and visibility conditions govern the migration intensity over the open sea (for details see Hüppop et al. 2006).

Birds generally try to fly at altitudes at which their energy costs are lowest (Bellrose 1967, Bruderer and Liechti 1995), which usually means that the flight altitude of low migrating birds is lower offshore than at the coast or inland (e.g. Krüger and Garthe 2001, Hüppop et al. 2004). One possible factor is that tailwinds at low altitudes are more favourable at sea than over land. Moreover, adverse weather conditions lead to further reductions in flight altitude, so that many birds fly low over the water surface (Hüppop et al. 2004).

The flight altitude distribution may also be affected by the range of species involved. Many diurnally migrating species of waterfowl and seabirds migrate mostly at very low altitudes (Krüger and Garthe 2001, Dierschke and Daniels 2003, Hüppop et al. 2004), while arctic waders fly at very high altitudes, at least when migrating to their breeding grounds in May (Green 2003). Altitude distribution differences between nights may therefore be due to a different range of species. In the long run however, the effect of this phenomenon on the overall picture is probably marginal.

Reverse migration often occurs in connection with changing weather conditions, e.g. when severe weather forces the birds to turn back (namely in spring) or when good weather periods "encourage" the birds to return to favourite stopover sites in autumn migration. In this context, reverse migration can be explained as an energy-saving strategy (Wikelski et al. 2003). Migrants may also be drifted by heavy gales (e.g. Alerstam 1990). Whatsoever, with respect to the planned offshore wind turbines return migration means that at least some birds have to pass the "dangerous zones" more than once per migration season, thus increasing the risk of collision for the individual.

9.4.3   Spatial Distribution

Reliable information about the spatial distribution of bird migration over the North Sea is, besides collision rate, probably the largest issue in assessing the impact of offshore wind farms on birds. Together with visual and radar investigation, analyses of ringing recoveries can provide at least a rough impression (Stolzenbach et al. in press).

Military surveillance radar data confirmed the findings by our visual and ship radar observations that migratory intensity decreases with increasing distance from the coast. However, the data also showed that there is considerable broad-front migration over the open sea, including birds obviously heading to or from southern Norway and eastern England, and many days with reverse migration (Hüppop et al. 2004).

Short term recoveries of ringed birds might indicate migration routes. Stolzenbach et al. (in press) analysed such recoveries for eleven species ringed in Norway, Sweden, Denmark and Germany, to reconstruct their flight routes over the southeastern North Sea. In autumn, broad-front migration towards the southwest seems to be the prevailing movement of, for example, Swedish, Danish and northern German dunnocks *Prunella modularis*, robins, garden warblers *Sylvia borin*, song thrushes, common redstarts *Phoenicurus phoenicurus* and pied flycatchers *Ficedula hypoleuca*. For species such as willow warbler *Phylloscopus trochilus* and black-headed gull however, the main direction is probably more south-southwest. The second important movement seems to be a broad-front migration from southern Norway to the south, in which the Norwegian and probably parts of the Swedish populations of e.g. dunnock, robin, garden warbler, willow warbler and probably dunlin are involved, whereas, for example, pied flycatchers obviously avoid long sea crossings. A few species, such as blackbird and robin, also seem to fly directly west from Denmark and Schleswig-Holstein to Britain. In spring, reverse migration routes with obviously stringer coastal orientation can be found. Some species appear to avoid crossing the German Bight, and migrate along the coastline (e.g. black-headed gulls, adult dunlins and dunnocks in spring). However, despite more than 100 years of bird ringing, for most species the data is still too poor for detailed reconstructions of their migration routes and the results have to be considered with extreme caution. Nevertheless they confirm general assumptions based on other methods at the species level.

9.4.4 Collisions

Bird strikes at sea are most frequently observed at oil and gas rigs with their flares (Sage 1979, Hope Jones 1980) and offshore marine research facilities (Müller 1981). The extent to which avian migrants are likely to interact with the projected offshore wind farms is currently difficult to estimate (Hüppop et al. 2004, Desholm and Kahlert 2005). Especially under weather conditions with poor visibility, illuminated objects can attract nocturnal migrants in large numbers (e.g. Schmiedel 2001).

Offshore wind turbines have to be illuminated for reasons of ship and aircraft safety, too. Collision is likely to be even more pronounced at sea than on land, as there are no suitable resting places at sea for terrestrial birds, especially during nights with dense migration traffic and adverse weather conditions. There are usually only a few such nights per migration period. Over half the bird cadavers collected on FINO 1 were found in just two nights. Our results clearly show that the mass of birds collected had collided with the structure, and in only a few cases could starvation not be ruled out entirely. Since most birds probably fell into the sea or were taken by gulls, the actual total of collisions is presumably many times higher.

The consequences of barrier effects on flight energetics are largely unknown, but are the subject of current research projects. The effect that collision mortality or higher flight costs will have on the population level is extremely difficult to predict, and depends on the life history and population status of the respective species. Dierschke et al. (2003) assumed that an increase of the existing adult mortality rate by 0.5 - 5 %, depending on the individual species, seems to be acceptable for the 250 bird species regularly migrating across the German maritime area. Any greater loss would have to be classed as "considerable impact". Because the area covered by a bird population during migration usually crosses borders, an international approach is necessary to assess impacts on populations of migrating birds. Meanwhile, M. Rebke et al. (in prep.) have developed much more detailed Leslie-matrix models for selected key-species, which will soon be available for predictions for the different scenarios. Further, collision models for different scenarios, including different sizes of wind farms and turbines as well as their arrangement within the farm, are under development.

9.5   Conclusions

Our findings confirm that large numbers of diurnal and nocturnal migrants cross the German Bight, with considerable variation of migration intensity, time, altitude and species, depending on season and weather conditions. This variability makes precise analyses, even more serious predictions, very difficult and further investigations necessary. Dierschke (2003) estimated from systematic visual observations that in 18 species, significant proportions (> 1 %) of the respective bio-geographical population pass Helgoland during migration, including more than 10 % of red-throated divers, pink-footed geese, greylag geese, brent geese and little gulls. Large numbers of nocturnally migrating birds of unknown species also cross the German Bight. Almost half the birds fly at "dangerous" altitudes, and the considerable reverse migration increases the risk of collision. Normally, migrating birds seem to avoid obstacles, even at night (Isselbächer and Isselbächer 2001, Schmiedel 2001, Desholm and Kahlert 2005), which diminishes collision risk, but increases flight costs. But at poor visibility caused by drizzle and mist, terrestrial birds in particular are attracted by illuminated offshore obstacles. Disoriented birds flew around the platform repeatedly, increasing both their risk of collision and their energy consumption.

In a few nights a year, a large number of avian interactions at offshore plants can be expected, especially in view of the planned number and extent of projected wind farms. Previous studies have been able to examine diurnal collisions only in good weather conditions, which are not predominant, or refer only such to large species such as geese and ducks, although smaller species are most frequently involved in collisions.

Despite the knowledge gaps, several mitigation measures can be recommended:

- Abandonment of plans for wind farms in zones with dense migration, e.g. in nearshore areas or along "migration corridors";
- alignment of turbines in rows parallel to the main migratory direction;
- several kilometre-wide free migration corridors between wind farms;

- no construction of wind farms between e.g. resting and foraging areas;
- shut-down of turbines at nights with bad weather/visibility and high migration intensity;
- refraining from large-scale continuous illumination;
- measures to make wind turbines generally more recognisable to birds.

In particular, the penultimate of these measures will require appropriate experiments with the brightness and colour of wind farm illumination, to minimise collision rates. Perhaps the most effective solution would be lighting adjusted to the weather conditions, e.g. flash-light with long intervals, instead of continuous light in fog and drizzle. During the very few nights in which a high frequency of bird strikes is expected, with predicted poor weather and high migration intensity, a shut-down of turbines and adjustment of rotor blades to minimise their surfaces relative to the main direction of migration could help reduce collisions.

Our findings also indicate that a combination of methods is necessary to describe the complex patterns of migration over the sea. However, even with virtually non-stop recording, as on FINO 1, the wide variation in bird migration and in weather (together with its effect on the former) lead to an insufficient number of samples per weather situation. The funding of further research in the follow-up project FINOBIRD (financed by the German Federal Ministry for the Environment, Nature Conservation and Nuclear Safety, grant no. 0329983) is a response to this problem. The recordings are to be continued with the aim of refining the results presented here.

Furthermore, we plan to develop a model to "forecast" bird migration over the German Bight with the aid of weather forecasts, for example to establish a basis for mitigation measures. However, as long as no investigations at existing wind farms are carried out to provide reliable data on collisions and avoidance behaviour, the actual scale of these problems will remain a matter of speculation. We expect more information on avoidance behaviour and collisions with the construction of a pilot wind farm close to the FINO 1 platform, which will be in 2007, at the earliest.

We are grateful to all those who helped with this comprehensive project, which was funded by The German Federal Ministry for the Environment, Nature Conservation and Nuclear Safety, grant no. 0327526. ### shortened and linguistically polished an earlier draft. Finally we thank Julia Köller and her colleagues for their patience with the authors and for editing this book.

References

Alerstam T (1990) Bird migration. Cambridge: Cambridge University Press

Bellrose FC (1967) The distribution of nocturnal migrants in the air space. Auk 88:397-424

Bruderer B (1997a) The study of bird migration by radar. Part 1: Technical basis. Naturwissenschaften 84:1-8

Bruderer B (1997b) The study of bird migration by radar. Part 2: Major achievements. Naturwissenschaften 84:45-54

Bruderer B, Jacquat B (1972) Zur Bestimmung von Flügelschlagfrequenzen tag- und nachtziehender Vogelarten mit Radar. Ornithol Beob 69:189-206

Bruderer B, Liechti F (1995) Variation in density and height distribution of nocturnal migration in the south of Israel. Israel J Zool 41:447-487

Bruderer B, Liechti F (2004) Welcher Anteil ziehender Vögel fliegt im Höhenbereich von Windturbinen. Ornithol. Beob. 101:327-335

Bruderer B, Steuri T, Baumgartner M (1995) Shortrange high-precision surveillance of nocturnal migration and tracking of single targets. Isr J Zool 41:207-220

Cooper B, Day R, Ritchie R, Cranor C. (1991) An improved marine radar system for studies of bird migration. J Field Ornithol 62:367-377

Dahlke C (2002) Genehmigungsverfahren von Offshore-Windenergieanlagen nach der Seeanlagenverordnung. Natur Recht 8:472-479

Desholm M, Kahlert J (2005) Avian collision risk at an offshore wind farm. Biol Lett 1:296-298

Dierschke V (2001) Vogelzug und Hochseevögel in den Außenbereichen der Deutschen Bucht (südöstliche Nordsee) in den Monaten Mai bis August. Corax 18:281-290

Dierschke V (2003) Quantitative Erfassung des Vogelzugs während der Hellphase bei Helgoland. Corax 19, Sonderheft 2:27-34

Dierschke V, Daniels J-P (2003) Zur Flughöhe ziehender See-, Küsten- und Greifvögel im Seegebiet um Helgoland. Corax 19: Sonderheft 2:35-41

Dierschke, J, Dierschke, V, Krüger, T (2005) Anleitung zur Planbeobachtung des Vogelzugs über dem Meer ('Seawatching'). Seevögel 26:2-13

Dierschke, V, Hüppop, O, Garthe, S (2003) Populationsbiologische Schwellen der Unzulässigkeit für Beeinträchtigungen der Meeresumwelt am Beispiel der in der deutschen Nord- und Ostsee vorkommenden Vogelarten. Seevögel 24:61-72

Drost R (1928) Über die Errichtung und Besetzung von Hilfsbeobachtungsstellen der Vogelwarte Helgoland. J Ornithol 76:471-472

Eastwood E (1967) Radar ornithology. London: Methuen

Fransson T (1998) Patterns of migratory fuelling in whitethroats *Sylvia communis* in relation to departure. J Avian Biol 29:569-573

Friebe T (1998) Vogelzugbeobachtungen mit Hilfe der Radargeräte des Radarführungsdienstes der Deutschen Luftwaffe. Vogel Luftverkehr 18:23-30

Gätke H (1891) Die Vogelwarte Helgoland. Braunschweig: Meyer

Green M (2003) Flight strategies in migrating birds: when and how to fly. PhD thesis, Lund University

Gruber S, Nehls G (2003) Charakterisierung des offshore Vogelzuges vor Sylt mittels schiffsgestützer Radaruntersuchungen. Vogelkdl Ber Niedersachs 35:151-156

Hope Jones P (1980) The effect on birds of a North Sea gas flare. Br Birds 73:547-555

Hüppop K, Hüppop O (2004) Atlas zur Vogelberingung auf Helgoland. Teil 2: Phänologie im Fanggarten, 1961 bis 2000. Vogelwarte 42:285-343

Hüppop O, Dierschke J, Wendeln H (2004) Zugvögel und Offshore-Windkraftanlagen: Konflikte und Lösungen. Ber. Vogelschutz 41:127-218

Hüppop O, Dierschke J, Exo K-M, Fredrich E, Hill, R (2006) Bird migration studies and potential collision risk with offshore wind turbines. Ibis 148:90-109

Hüppop O, Exo K-M, Garthe S (2002) Empfehlungen für projektbezogene Untersuchungen möglicher bau- und betriebsbedingter Auswirkungen von Offshore-Windenergieanlagen auf Vögel. Ber Vogelschutz 39:77-94

Isselbächer K, Isselbächer T (2001) Windenergieanlagen. In Richarz K, Bezzel E, Hormann M (eds) Taschenbuch für Vogelschutz: 128-142. Wiebelsheim: Aula

Jacobs J (1974) Quantitative measurement of food selection. Oecologia 14:413-417

Jellmann J, Vauk G (1978) Untersuchungen zum Verlauf des Frühjahrszuges über der Deutschen Bucht nach Radarstudien und Fang- und Beobachtungsergebnissen auf Helgoland. J. Ornithol. 119:265-286

Krüger T, Garthe S (2001) Flight altitudes of coastal birds in relation to wind direction and speed. Atlantic Seabirds 3:203-216

Merck T, von Nordheim, H (2000) Technische Eingriffe in marine Lebensräume. BfN-Skripten 29. Bonn-Bad Godesberg

Müller HH (1981) Vogelschlag in einer starken Zugnacht auf der Offshore-Forschungsplattform 'Nordsee' im Oktober 1979. Seevögel 2:33-37

Sage B (1979) Flare up over North Sea birds. New Scientist 82:464-466

Schmiedel J (2001) Auswirkungen künstlicher Beleuchtung auf die Tierwelt – ein Überblick. Schriftenr. Landschaftspflege Naturschutz 67:19-51

Stolzenbach F, Ballasus H, Hüppop O, Rebke, M (in press) Rekonstruktion von Vogelzugwegen über die südöstliche Nordsee anhand von Ringfunden. Acta ornithoecologica

Sutter E (1957) Radar als Hilfsmittel der Vogelzugforschung. Ornithol Beob 54:70-95

Walter G, Todeskino D (2005) Zur Richtung und Höhenverteilung des Vogelzuges im Bereich Nordergründe (Wesermündung) auf der Grundlage von Radaruntersuchungen. Natur- und Umweltschutz 4:29-35

Wikelski M, Tarlow EM, Raim A, Diehl RH, Larkin RP, Visser GH (2003) Costs of migration in free-flying songbirds. Nature 423:704

Research on Resting and Breeding Birds

Background

A total of thirty-five species of sea bird occur in German waters. For many of these sea birds and waterfowl species, particularly the near-coastal areas of the German North and Baltic Seas up to water depths of approx. 30 m are of very great, and even vital, importance, as feeding, resting, moulting and wintering areas. The distribution of the sea birds in space and time is largely determined by season, weather conditions and food supply.

In the German Exclusive Economy Zone of the North Sea, the eastern German Bight has been identified as the area most suitable for a bird reserve in terms of extent and population, and has been proposed for that purpose under the EU Birds Directive. In the Baltic Sea, the establishment of protected status under EU law involves the Pomeranian Bight, which is of special ecological significance, particularly for wintering species.

The construction and operation of offshore wind-power turbines involves on the one hand the risk of collision for sea and resting birds with the turbines, and on the other the possibility of disturbance effects on resting, feeding and wintering of sea birds. Experience with wind-power turbines on land has shown that rapidly turning rotor blades can trigger flight reactions in birds, and that the plants can hence result in habitat loss. Dislocation and scaring-off effects are to be expected not only from wind-power turbines, but also from construction and supply vehicles. Particularly for sensitive species, it is to be assumed that they will avoid the area of a wind park, and that such area will therefore be lost as resting areas.

As for migratory birds, so too for resting and feeding birds which live on the high seas for lengthy periods, offshore wind-power turbines may constitute a barrier. The fragmentation effect of the plant areas may separate ecologically associated resting and feeding sites.

With regard to the analysis and assessment of possible consequences of the construction, operation and dismantling of offshore wind-power turbines required in the context of the Environmental Impact Assessment, the central question to be answered is whether the effects will reach such an intensity in practice that endangerment of the marine environment would become possible, and authorisation of the proposed wind park would have to be denied. For this purpose, the following correlations of effects must essentially be considered:

- Short-term loss of habitat due to a temporary scaring-off effect during the construction phase, or to maintenance work (particularly due to ship and air traffic);
- Long-term loss of habitat due to the scaring-off effect of the plants;

- Fragmentation of ecologically associated units, such as resting and feeding areas (barrier effect);
- Direct loss of individuals due to collisions (especially for nocturnal species).

Another issue being discussed is the extent to which the introduction of hard substrata, e.g., due to the construction of the foundations of the plants, could result in a change of the species spectrum of the benthos and fish communities in the area of the wind park, and thus also change the feeding areas of benthos and fish-eating bird species as a result.

For an assessment of the conflict potential of these correlations of effects, both the occurrence of sea and resting birds in terms of time and space, and details regarding their general behaviour (feeding, effects of the weather) and their behaviour towards offshore wind-power turbines, or towards construction and supply vehicles (flight distances, circumnavigation movements, collision risk) must be ascertained.

The research projects carried out in the context of ecological accompanying research on marine birds aims primarily at closing existing knowledge gaps on offshore habitats in order to be able to observe the conflict potential between offshore wind power use and important resting bird occurrences at sea more objectively. In addition to the ascertainment of the extensive distribution and description of species and species group patterns in time and spatial terms, so-called disturbance experiments, by means of which the flight distances of some species could better be assessed, were central.

10 Possible Conflicts between Offshore Wind Farms and Seabirds in the German Sectors of North Sea and Baltic Sea

Volker Dierschke, Stefan Garthe, Bettina Mendel

10.1 Introduction

Since the beginning of the discussion about offshore wind farms, the effects of wind turbines on seabirds have been among the most prominent aspects of possible impact on the marine environment. Habitat loss due to displacement and avoidance as well as by the introduction of hard substrate into a soft bottom environment has been most prominently in the focus of discussion, but habitat fragmentation caused by detours flown, and to a lesser extent also increased mortality due to fatal collisions have also been considered (see reviews by Garthe 2000, Noer et al. 2000, Exo et al. 2002). While first results from operating offshore wind farms in Denmark and Sweden have confirmed a number of assumptions for some, but not all seabird species (e.g. Petersen et al. 2004, Pettersson 2005; summarized by Dierschke and Garthe 2005), research in German marine areas is currently concentrated on assessing possible impacts by I) surveying seabird distribution and II) developing assessment methods. In this respect, the seabird research included in the MINOS[1] project was of considerable importance (Garthe et al. 2004).

A total of 35 seabird species have been identified as living regularly in the German territorial waters and the Exclusive Economic Zone (EEZ) (Garthe et al. 2003a). These include birds breeding in coastal colonies and foraging offshore as well as birds wintering, moulting and/or staging during migration at sea. Ship-based transect counts have shown the distribution of most of these species (Durinck et al. 1994, Skov et al. 1995, Mitschke et al. 2001, Garthe et al. 2003b), but incomplete knowledge still existed for various areas and/or seasons. The aim of the MINOS project was therefore to fill these gaps by additional ship-based mapping, but also – for the first time in Germany – to obtain large-scale snapshots of

---

[1] MINOS - Marine Warmblüter in Nord- und Ostsee (Marine warm-blooded animals in the North and Baltic Seas: Foundations for assessment of offshore wind farms)

seabird distributions by aerial surveys. This should help to identify areas suitable for wind farms, and areas that need to be protected by law (e.g. according to the EU Birds Directive). In order to assess the importance of marine areas for seabirds, two methods have been developed I) the Wind Farm Sensitivity Index (Garthe and Hüppop 2004) and II) the application of threshold levels (Dierschke et al. 2003).

10.2   Distribution of Seabirds

Some examples are selected in order to show I) the type of results which can be obtained from aerial and ship-based surveys and II) the seasonal variation of seabird distribution. The complete results of recent seabird mapping in German waters are available from Garthe et al. (2003b, 2004).

Red-throated Diver (*Gavia stellata*) and **Black-throated Diver** (*Gavia arctica*): The two diver species are difficult to separate by the survey methods applied at sea, especially when performing aerial surveys. In winter, the proportion of Red-throated Divers is 95 % in the German North Sea and 46 % in the German Baltic Sea (Mitschke et al. 2001, Garthe et al. 2003b). Considerable numbers of divers are present in German waters from November to April, but their distribution changes during winter and spring. In winter, their occurrence in the North Sea is concentrated in a 20 km (Lower Saxony) to 60 km wide strip (Schleswig-Holstein) along the coast (Fig. 1), whereas divers are distributed more evenly in most parts of the Baltic Sea (Fig. 2). In springtime, divers move both to the north and to the east, leading to concentrations west of the North Frisian Islands (Fig. 3) and east of the island of Rügen (Fig. 4). From these two areas, the birds seem to depart to their breeding areas in northern Eurasia.

Common Scoter (*Melanitta nigra*): The Common Scoter is one of the most abundant seabirds in German waters. Whereas these birds are restricted to coastal areas in the North Sea, they are present further offshore in the Baltic Sea. Throughout the year, the most important concentrations occur in the shallow areas of the Oderbank (Pomeranian Bay), which is used not only for wintering (Fig. 5) and migratory staging (Fig. 6), but also during summer for moult (Fig. 7, Sonntag et al. 2004). In winter and spring, the western parts of the Baltic Sea are also of importance (Fig. 5 and 6).

Common Guillemot (*Uria aalge*): Whereas in summer, Guillemots are mainly restricted to the sea area around Helgoland (Fig. 8, Dierschke et al. 2004), they are dispersed throughout most of the German EEZ in wintertime (Fig. 11). Snapshots by aerial surveys in March and April 2003 show how (probably British) auks disappear and how breeding birds concentrate around Helgoland at this time of the year (Fig. 9 and 10).

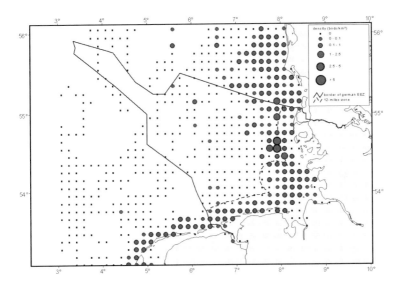

Fig. 1. Winter distribution of divers in the German Bight according to ship-based surveys (only data up to seastate 4, 1 November to 29 February, 1980 - 2003)

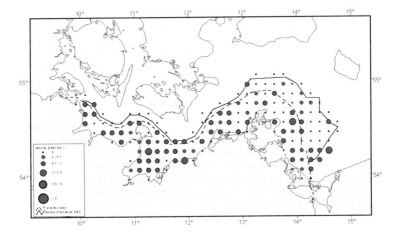

Fig. 2. December distribution of divers in the south-western Baltic Sea according to aerial surveys (10 - 12 December 2002, 7 - 9 December 2003)

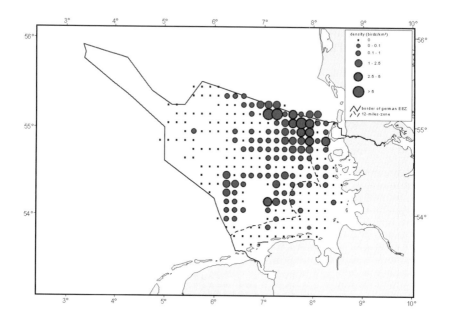

Fig. 3. April distribution of divers in the German Bight according to aerial surveys (17 - 24 April 2003)

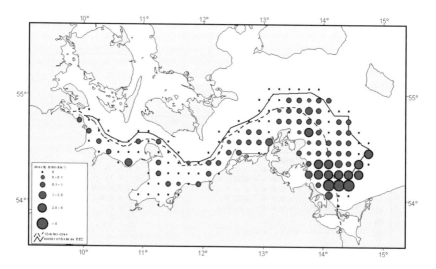

Fig. 4. April distribution of divers in the south-western Baltic Sea according to aerial surveys (12 - 14 April 2003)

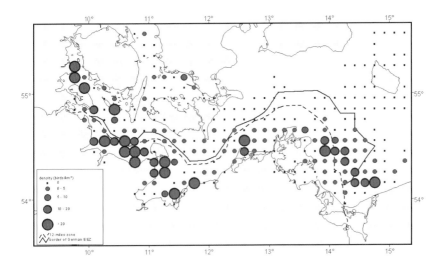

Fig. 5. Winter distribution of Common Scoters in the south-western Baltic Sea according to ship-based surveys (only data up to seastate 4, 1 December to 29 February, 1980 - 2003)

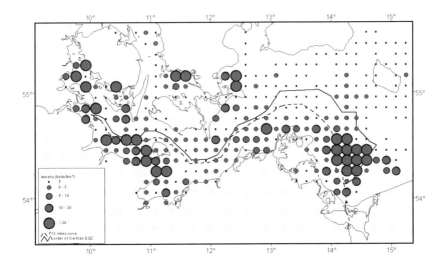

Fig. 6. Spring distribution of Common Scoters in the south-western Baltic Sea according to ship-based surveys (only data up to seastate 4, 1 March to 31 May, 1980 - 2003)

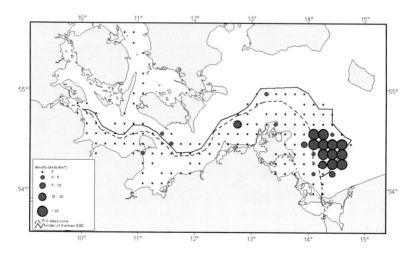

Fig. 7. Summer distribution of Common Scoters in the south-western Baltic Sea according to ship-based surveys (only data up to seastate 4, 1 June to 30 September, 1980 - 2003)

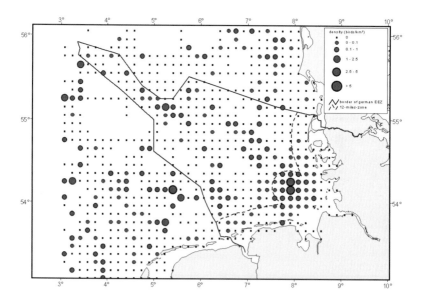

Fig. 8. Summer distribution of Common Guillemots in the German Bight according to ship-based surveys (only data up to seastate 4, 16 April to 30 June, 1980 - 2003)

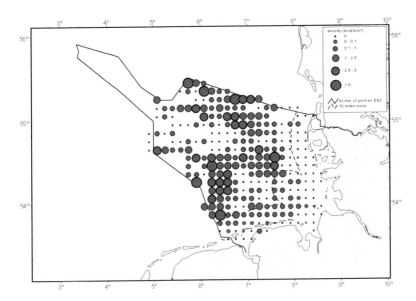

Fig. 9. March distribution of Common Guillemots and Razorbills in the German Bight according to aerial surveys (13 - 20 March 2003)

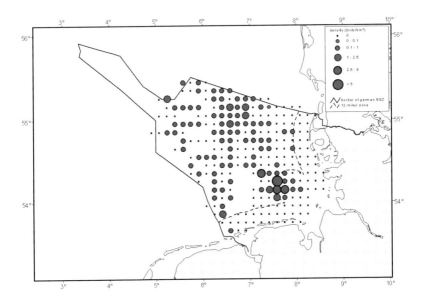

Fig. 10. April distribution of Common Guillemots and Razorbills in the German Bight according to aerial surveys (17 - 24 April 2003)

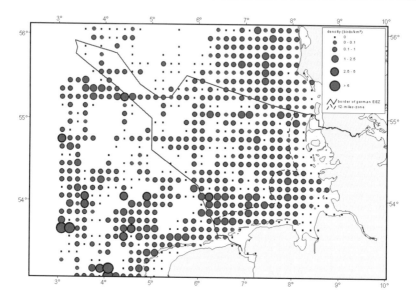

Fig. 11. Winter distribution of Common Guillemots in the German Bight according to ship-based surveys (only data up to seastate 4, 1 October to 29 February, 1980 - 2003)

10.3   Assessing the Vulnerability of Seabirds to Offshore Wind Farms

During the MINOS project, an index was developed to obtain species-specific estimates of vulnerability of seabirds to offshore wind farms (for details see Garthe and Hüppop 2004). Nine factors which are thought to be relevant in the context of disturbance and collision risk were selected: (1) flight manoeuvrability, (2) flight altitude, (3) percentage of time flying, (4) nocturnal flight activity, (5) disturbance by ship and helicopter traffic, (6) flexibility in habitat use, (7) bio-geographic population size, (8) adult survival rate and (9) European threat and conservation status. Hence, these factors represent flight behaviour (1 - 4), general behaviour (5 - 6) and status (7 - 9). For each species, each factor was scored from 1 (low vulnerability) to 5 (high vulnerability), resulting in Species Sensitivity Indices (SSI). While five factors were assessed by research data, the other four were estimated according to the experience from at-sea observations.

For those species occurring in the German sector of the North Sea, the highest SSI values were calculated for the Red-throated Diver and the Black-throated Diver, the Velvet Scoter (*Melanitta fusca*) and the Sandwich Tern (*Sterna sandvicensis*), whereas vulnerability according to SSI was the lowest for a number of gull species and for the Northern Fulmar (*Fulmarus glacialis*) (Table 1).

Table 1. Species Sensitivity Index (SSI) values for the 26 seabird species occurring in the German sector of the North Sea. For scores of the nine vulnerability factors see Garthe and Hüppop (2004)

Species	SSI
Black-throated Diver (*Gavia arctica*)	44.0
Red-throated Diver (*Gavia stellata*)	43.3
Velvet Scoter (*Melanitta fusca*)	27.0
Sandwich Tern (*Sterna sandvicensis*)	25.0
Great Cormorant (*Phalacrocorax carbo*)	23.3
Common Eider (*Somateria mollissima*)	20.4
Great Crested Grebe (*Podiceps cristatus*)	19.3
Red-necked Grebe (*Podiceps grisegena*)	18.7
Great Black-backed Gull (*Larus marinus*)	18.3
Black Tern (*Chlidonias niger*)	17.5
Common Scoter (*Melanitta nigra*)	16.9
Northern Gannet (*Sula bassana*)	16.5
Razorbill (*Alca torda*)	15.8
Atlantic Puffin (*Fratercula arctica*)	15.0
Common Tern (*Sterna hirundo*)	15.0
Lesser Black-backed Gull (*Larus fuscus*)	13.8
Arctic Tern (*Sterna paradisaea*)	13.3
Little Gull (*Hydrocoloeus minutus*)	12.8
Great Skua (*Stercorarius skua*)	12.4
Common Guillemot (*Uria aalge*)	12.0
Common Gull (*Larus canus*)	12.0
Herring Gull (*Larus argentatus*)	11.0
Arctic Skua (*Stercorarius parasiticus*)	10.0
Black-headed Gull (*Larus ridibundus*)	7.5
Black-legged Kittiwake (*Rissa tridactyla*)	7.5
Northern Fulmar (*Fulmarus glacialis*)	5.8

By multiplying the SSI values with the density of the respective species, the question of vulnerability can be transferred to given areas of sea. Summing up the resulting values of all species yields the area-specific Wind Farm Sensitivity Index (WSI). This procedure was applied to the German North Sea. It became obvious that especially the coastal areas of the south-eastern North Sea must be regarded as vulnerable to wind farms (Fig. 12). Threshold WSI levels were proposed for various levels of concern, further indicating parts of the German North Sea which are sensitive to disturbance from wind turbines (Fig. 13). In combination with the large-scale mapping of seabird densities, the WSI will be a valuable tool in order to consider the results of small-scale case studies (impact assessments for single wind farms) in a wider context.

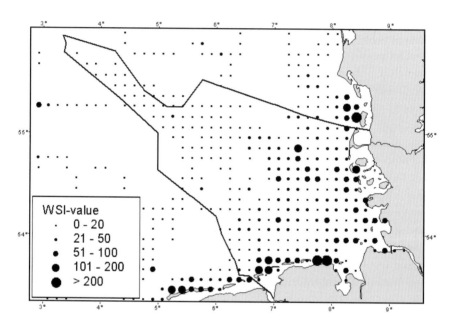

Fig. 12. Spatial distribution of the wind farm sensitivity index (WSI) values (all seabird species combined) in the south-eastern North Sea during springtime (March - May) 1993 - 2003. (Garthe and Hüppop 2004, Fig. 4)

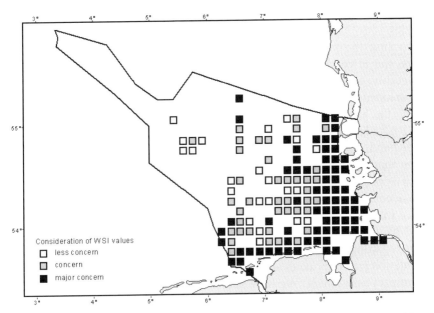

Fig. 13. Areas in the German sector of the North Sea, where wind energy utilization is considered to be of "no (less) concern", "concern" or "major concern". Areas not studied during at least one of the seasons are left blank. (Garthe and Hüppop 2004, Fig. 6)

10.4   Assessing the Possible Impact of Offshore Wind Farms on Seabirds

Effects of offshore wind farms on seabirds impact their population dynamics as soon as either their mortality rate or reproduction rate are affected to a degree that changes the population size. As it will be nearly impossible to detect a direct connection between effects of wind turbines and the parameters included in population dynamics, and because even indirect effects due to density-dependent processes acting among seabirds at sea are poorly known, simpler approaches are necessary. In addition to a methodology for impact assessment which combines the sensitivity of the seabirds occurring with the magnitude of the disturbing effects (Percival 2001), a simple way of applying threshold levels was proposed in the MINOS project (Dierschke et al. 2003).

Based on the commonly used criteria from the Ramsar Convention (Atkinson-Willes 1972), the occurrence of 1 % of a population was considered to be the level that indicates that a given area is important for a species.

Because the population size can be affected by several technical impacts, the effects of all these impacts must be considered cumulatively. Preferably, all impacts in the entire area used by a seabird species should be included in a risk analysis, with the entire population as the unit to which the magnitude of effects is to be referenced. However, on the one hand population sizes are in many cases not accurately known. On the other hand, in the case of offshore wind farms, it must be taken into account that in a competitive business and under different legislative scenarios it appears unrealistic that the commissioning of wind farms will take into account technical impacts by the wind farms in other countries, especially for areas far away from Germany – but which may nevertheless be used by the same population of the same seabird species. Therefore, at least regarding Germany the better known and easier to handle national population size (Garthe et al. in prep.) would seem to be much better suited for impact assessment. We have shown how the 1%-criterion might be applied in practise by giving examples for some seabird species in the German sector of the North Sea.

In the territorial waters and the Exclusive Economic Zone, a total of 26 offshore wind farms has been proposed, nine of which had been approved as by August 2005 (Fig. 14). The size of all proposed wind farms is 1,297 km², those approved cover 340 km². Seabird densities were calculated for all four seasons (winter, spring, summer, autumn) for seven areas that contain one or more wind farms each (Fig. 14).

In order to obtain an estimate of how many seabirds live in the wind farm areas, bird densities were multiplied by the size of the wind farms (Northern Gannet *Sula bassana*, Lesser Black-backed Gull *Larus fuscus*, Sandwich Tern). For those species which were observed to strongly avoid wind farms, with distances kept from turbines of 2 km or more (Red-throated Diver, Common Guillemot; Petersen 2005, Dierschke and Garthe 2005), a buffer zone of 2 km was included in the calculation[2]. Compared to the national population size of the respective species, the number of birds and the proportion of the national population either living in wind farms (e.g. foraging under the risk of collision) or loosing habitat are calculated. It is important to note that almost the same data sets have been used for calculating both national and regional numbers for wind farms.

---

[2]   all wind farms plus buffer zones cover 2,333 km², those wind farms with approval cover 757 km²; overlaps between wind farms have been considered

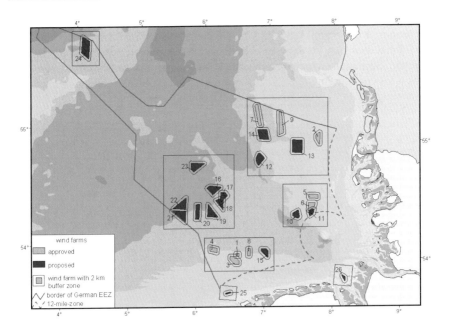

Fig. 14. Wind farms proposed and/or approved in the German sector of the North Sea by August 2005. Rectangles indicate the areas used for the calculation of seabird densities. Approved: 1 Borkum West (12 turbines), 2 Butendiek (80), 3 Borkum Riffgrund (77), 4 Borkum Riffgrund West (80), 5 Amrumbank West (80), 6 Nordsee Ost (80), 7 Sandbank 24 (80), 8 ENOVA Offshore Northsea Windpower (48), 9 DanTysk (80)

The Figs. 15 - 19 show the cumulative effects of proposed and/or approved offshore wind farms on seabirds in the German sector of the North Sea, expressed as the number of individuals and the respective proportions of the German populations affected by habitat loss (Red-throated Divers, Northern Gannets, Common Guillemots) or collision risk (Lesser Black-backed Gulls, Sandwich Terns). The numbers 1 - 9 refer to the wind farms approved by August 2005, in order of their date of approval (compare numbers in Fig. 14); the value for 26 represents the seabirds concerned if all proposed wind farms are constructed.

Table 2. Proportions of selected German seabird populations affected by approved and proposed offshore wind farms in the German sector of the North Sea. Note that habitat loss can be expected for Red-throated Diver, Northern Gannet and Common Guillemot, whereas in the other two species due to flights through wind farms a loss of foraging habitat and mortality from collisions cannot be excluded. * Including a 2 km buffer zone. National population sizes according to Garthe et al. (in prep.).

species	season (day/month)	national population size	birds affected (9 approved wind farms)	birds affected (26 proposed wind farms)
Red-throated Diver* (*Gavia stellata*)	winter (1/11-29/2)	5,000	35 (0.7%)	115 (2.3%)
	spring (1/3-15/5)	12,000	645 (5.4%)	1417 (11.8%)
Northern Gannet (*Sula bassana*)	summer (1/5-31/8)	1,400	16 (1.1%)	36 (2.6%)
	autumn (1/9-31/10)	2,700	18 (0.7%)	47 (1.7%)
Lesser Black-backed Gull (*Larus fuscus*)	summer (16/5-15/7)	46,000	855 (1.9%)	1446 (3.1%)
	autumn (16/7-31/10)	31,000	177 (0.6%)	550 (1.8%)
Sandwich Tern (*Sterna sandvicensis*)	summer (16/5-15/7)	3,000	7 (0.2%)	17 (0.6%)
	autumn (16/7-15/10)	2,000	2 (0.1%)	6 (0.3%)
Common Guillemot* (*Uria aalge*)	summer (16/4-30/6)	5,000	154 (3.1%)	557 (11.1%)
	winter (1/10-29/2)	33,000	752 (2.3%)	1899 (5.8%)

For **Red-throated Divers**, the German Bight is of importance especially as a spring staging site (Fig. 3, Garthe et al. 2004). Considerable numbers are present in March and April west of the Schleswig-Holstein coast, and many of them use the areas of sea designated for wind farms. Considering a 2 km buffer zone (which seems to be a conservative estimate, see below), no less than 11.8 % of the national springtime population would face habitat loss, if all proposed wind farms are in fact built and those wind farms already approved would take away habitat for 5.4 % of the birds (Table 2 and Fig. 15).

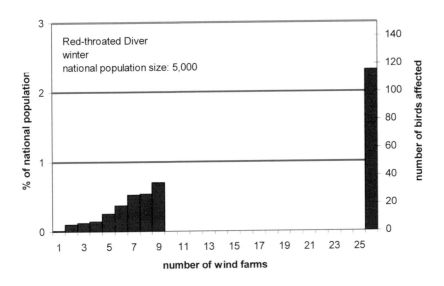

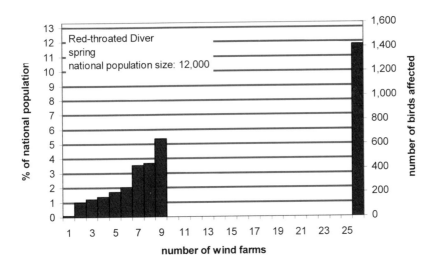

Fig. 15. Cumulative effects of proposed and/or approved offshore wind farms on Red-throated Divers in the German sector of the North Sea, expressed as the number of individuals and the respective proportion of the German population concerned from habitat loss

Taking into account that all divers have been displaced from the operating Horns Rev wind farm and decreased by 96.2 % at 0 - 2 km distance and 55.8 % at 2 - 4 km distance from the wind farm (Petersen 2005), the number of Red-throated Divers affected would be even larger: 1942 birds (all proposed wind farms, 16.2 % of the national population) or 868 birds (all approved wind farms, 7.2 %) would be displaced. It is feared that the high concentration of wind farms west of the island of Sylt will have considerable impact on staging divers, because the three already approved wind farms alone can be expected to displace at least 699 Red-throated Divers from their habitat. Due to turnover of individuals during the migration season, an even higher proportion can be expected. In addition, up to 2.3 % of the wintering population of 5,000 Red-throated Divers would loose their habitat (Table 2), and applying the avoidance measured in Horns Rev this proportion would be as high as 3.2 %. As movements within the winter quarter are not known and divers usually detour offshore wind farms (Christensen and Hounisen 2005), even more losses could be expected due to barrier effects and resulting habitat fragmentation (Dierschke and Garthe 2005).

Northern Gannets are distributed at low densities throughout the German Bight (Garthe et al. 2004). They were observed to detour the Horns Rev wind farm, but do not keep as far away from turbines as divers (Christensen and Hounisen 2005). For those wind farms already approved, a threshold level of 1 % of the population size is not reached, but all 26 proposed wind farms would affect up to 2.6 % of the summer population (Table 2 and Fig. 16).

In contrast to the other *Larus* species, **Lesser Black-backed Gulls** are typical foragers of offshore areas, and are spread throughout much of the German Bight in summer and autumn (Garthe et al. 2004, Garthe and Schwemmer 2005). This species was found to fly through wind farms at the coast as well as at sea (Everaert 2003, Christensen and Hounisen 2005), but it is unknown whether wind farm areas are used for foraging. As gulls generally seem to visit offshore wind farms, they appear to be affected by mortality due to collision rather than by habitat loss. Whatever the consequences for this species are, up to 3.1 % of the German population would be affected (Fig. 17).

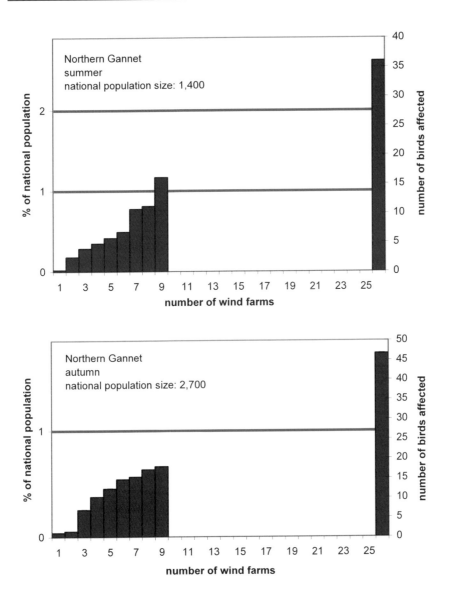

Fig. 16. Cumulative effects of proposed and/or approved offshore wind farms on Northern Gannet in the German sector of the North Sea, expressed as the number of individuals and the respective proportion of the German population concerned from habitat loss

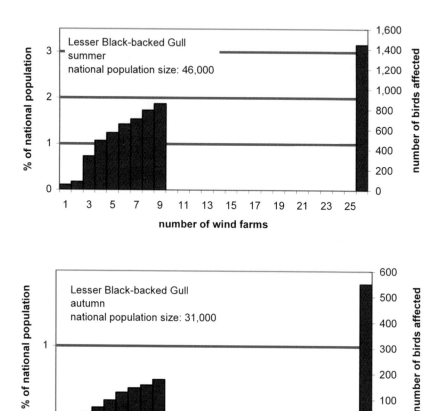

Fig. 17. Cumulative effects of proposed and/or approved offshore wind farms on Lesser Black-backed Gull in the German sector of the North Sea, expressed as the number of individuals and the respective proportion of the German population concerned from collision risk

Sandwich Terns are concentrated in a few large breeding colonies along the German North Sea coast, and mainly use only those parts of the sea for foraging located relatively close to their colonies (Garthe et al. 2004). Therefore, only a small proportion of the national summer and autumn populations would be affected by offshore wind farms (Table 2 and Fig. 18).

As Sandwich Terns were observed to fly into offshore wind farms (Christensen and Hounisen 2005), and the flight altitude of terns is usually lower than the rotor height of the turbines (e.g. Pettersson 2005), habitat loss and collision risk seem to be of minor importance.

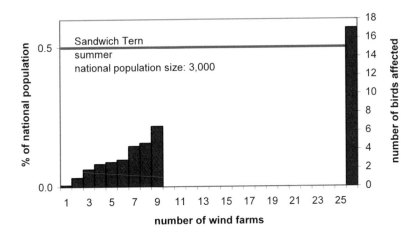

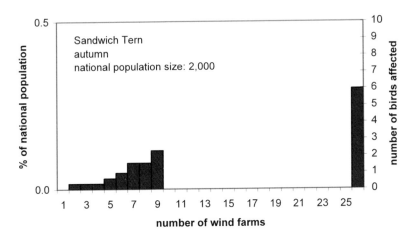

Fig. 18. Cumulative effects of proposed and/or approved offshore wind farms on Sandwich Tern in the German sector of the North Sea, expressed as the number of individuals and the respective proportion of the German population concerned from collision risk

In winter, **Common Guillemots** occur throughout most parts of the German Bight, but in summer, their distribution is more patchy, with a concentration in the surroundings of the only German breeding colony, on Helgoland (Fig. 8 and 9, Garthe et al. 2004, Dierschke et al. 2004).

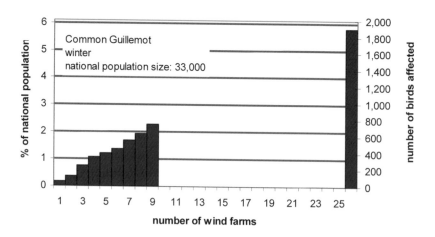

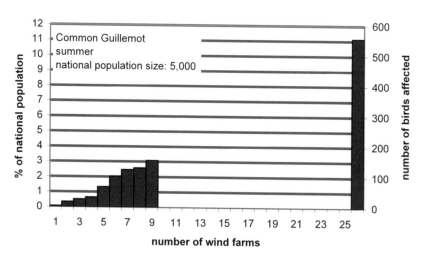

Fig. 19. Cumulative effects of proposed and/or approved offshore wind farms on Common Guillemot in the German sector of the North Sea, expressed as the number of individuals and the respective proportion of the German population concerned from habitat loss

Studies at Horns Rev found that auks (Common Guillemots and Razor-bills (*Alca torda*)) strictly avoid the wind farm and the 2 km zone around it and also do not fly through the turbine rows (Petersen 2005, Christensen and Hounisen 2005). Including a 2 km buffer zone, the wind farms already approved would lead to habitat loss for 3.1 % of the summer population and 2.3 % of the winter population. If all proposed wind farms are built, up to 11.1 % of the German population would be affected (Table 2 and Fig. 19).

10.5 Conclusion

In conclusion, it is obvious that depending on avoidance behaviour and collision risk as well as on the proportions of populations affected, the impact of offshore wind farms in the German sector of the North Sea on seabird populations differs considerably. The example of Red-throated Divers and Common Guillemots shows that large parts of the German Bight would be excluded from use by these species. This has to be taken into account in the process of commissioning by authorities and should lead to the application of threshold levels in order to select wind farm sites which have least impact on seabird populations. The above examples demonstrate that in the assessment of the effects of a single wind farm, the habitat loss in addition to the habitat already lost before due to other wind farms must be considered. Therefore, this underscores the need for a cumulative approach when assessing impacts on seabird populations.

References

Atkinson-Willes GL (1972) The international wildfowl censuses as a basis for wetland evaluation and hunting rationalization. In: Carp E (ed), Proc Int Conf Conserv of Wetland and Waterfowl, Ramsar 1971:87-119

Christensen TK, Hounisen JP (2005) Investigations of migratory birds during operation of Horns Rev offshore wind farm. Annual status report 2004. NERI Report commissioned by Elsam Engineering A/S

Dierschke V, Garthe S (2005) Review of the literature of ecological studies on seabirds at offshore wind farms. F+E project „Internationaler Erfahrungsaustausch zur ökologischen Begleitforschung von Offshore-Windenergieanlagen in Nord- und Ostsee", final report part B. Forschungs- und Technologiezentrum Westküste, Büsum

Dierschke V, Hüppop O, Garthe S (2003) Populationsbiologische Schwellen der Unzulässigkeit für Beeinträchtigungen der Meeresumwelt am Beispiel der in der deutschen Nord- und Ostsee vorkommenden Vogelarten. Seevögel 24:61-72

Dierschke V, Garthe S, Markones N (2004) Aktionsradien Helgoländer Dreizehenmöwen *Rissa tridactyla* und Trottellummen *Uria aalge* während der Aufzuchtphase. Vogelwelt 125:11-19

Durinck J, Skov H, Jensen FP, Pihl S (1994) Important marine areas for wintering birds in the Baltic Sea. Ornis Consult report 1994, Copenhagen

Everaert J (2003) Windturbines en vogels in Vlaanderen: voorlopige onderzoeksresultaten en aanbevelingen. Natuur Oriolus 69:145-155

Exo K-M, Hüppop O, Garthe S (2002) Offshore-Windenergieanlagen und Vogelschutz. Seevögel 23:83-95

Garthe S (2000) Mögliche Auswirkungen von Offshore-Windenergieanlagen auf See- und Wasservögel der deutschen Nord- und Ostsee. In: Merck T, von Nordheim H (eds.) Technische Eingriffe in marine Lebensräume: 113-119. BfN-Skripten 29, Bundesamt für Naturschutz, Bonn-Bad Godesberg

Garthe S, Hüppop O (2004) Scaling possible adverse effects of marine wind farms on seabirds: developing and applying a vulnerability index. J Appl Ecol 41:724-734

Garthe S, Schwemmer P (2005) Seabirds at Sea – Untersuchungen in den deutschen Meeresgebieten. Vogelwelt 126:67-74

Garthe S, Krüger T, Kubetzki U, Weichler T (2003a) Monitoring von Seevögeln auf See: Gegenwärtiger Stand und Perspektiven. Ber Landesamtes Umweltsch Sachsen-Anhalt, Sonderheft 1/2003:62-64

Garthe S, Ullrich N, Weichler T, Dierschke V, Kubetzki U, Kotzerka J, Krüger T, Sonntag N, Helbig AJ (2003b) See- und Wasservögel der deutschen Ostsee. Verbreitung, Gefährdung und Schutz. Bundesamt für Naturschutz, Bonn-Bad Godesberg

Garthe S, Dierschke V, Weichler T Schwemmer P (2004) Teilprojekt 5 - Rastvogelvorkommen und Offshore-Windkraftnutzung: Analyse des Konfliktpotenzials für die deutsche Nord- und Ostsee. In: Kellermann A et al. (eds): Marine Warmblüter in Nord- und Ostsee: Grundlagen zur Bewertung von Windkraftanlagen im Offshore-Bereich. Endbericht

Garthe S, Schwemmer P, Sonntag N. Dierschke V. (in prep.) Estimation of numbers of seabirds in the German North Sea throughout the yearly cycle and their biogeographic importance

Mitschke A, Garthe S Hüppop O (2001) Erfassung der Verbreitung, Häufigkeiten und Wanderungen von See- und Wasservögeln in der deutschen Nordsee. BfN-Skripten 34, Bonn-Bad Godesberg

Noer H, Christensen TK, Clausager I, Petersen IK (2000) Effects on birds of an offshore wind park at Horns Rev: Environmental impact assessment. NERI Report

Percival SM (2001) Assessment of the effects of offshore wind farms on birds. ETSU Report W/13/00565/REP, DTI/PubURN01/1434

Petersen IK (2005) Bird numbers and distribution in the Horns Rev offshore wind farm area. Annual status report 2004. NERI Report commissioned by Elsam Engineering A/S

Petersen IK, Clausager I, Christensen TK (2004) Bird numbers and distribution in the Horns Rev offshore wind farm area. Annual status report 2003. NERI Report commissioned by Elsam Engineering A/S

Pettersson J (2005) The impact of offshore wind farms on bird life in southern Kalmar Sound, Sweden. Report requested by Swedish Energy Agency

Skov H, Durinck J, Leopold MF, Tasker ML (1995) Important bird areas for seabirds in the North Sea including the Channel and the Kattegat. BirdLife International, Cambridge

Sonntag N, Engelhard O, Garthe S (2004): Sommer- und Mauservorkommen von Trauerenten *Melanitta nigra* und Samtenten *M. fusca* auf der Oderbank (südliche Ostsee). Vogelwelt 125:77-82

Research on Fish

Background

The North Sea provides a habitat for approx. 250 fish species, and is among the most productive fishing waters in existence. The fish fauna of the Baltic Sea is somewhat species-poorer than that of the North Sea, with a total of 144 species, of which 97 species are sea fish, seven are migrating fish and 40 are freshwater fish. While in the North Sea, the distance from the coast seems to be the main factor determining the composition of the species community, in the Baltic Sea it is the salt content gradient.

The fish populations in the North and Baltic Seas are subject to heavy burdens which are primarily due to the increasing use pressure on these bodies of water. Intensive fishing is decimating the fish populations and destroying habitats. Pollutant immission and eutrophication also have a major effect on fish populations. This involves in particular pollution by nitrates, phosphates, heavy metals and chlorinated hydrocarbons. In addition, there is noise pollution, particularly due to raw material extraction and explosive devices used in military exercise areas.

In connection with the construction and operation of offshore wind-power turbines, at present a variety of possible impairments of the fish fauna are being discussed. Particularly low-frequency sound can hurt fish physically, or induce flight reactions. Moreover, during the construction phase, sediment disturbance and strips of turbidity in the immediate proximity of the foundations could result in damage to the fish spawn or clogging of their gill apparatus. Fish could, via their air bladders, experience concussions and vibrations caused by the installation work for the foundations.

Overall, impairments of fish fauna might arise from the following correlations of effects, due to the construction and operation of offshore wind-power turbines:

- damage or dislocation of fish due to strips of turbidity during the construction phase;
- damage or dislocation of fish due to vibrations and noise during construction and operation of the plants;
- barrier effects within and outside the wind park due to electromagnetic fields;
- warming of the sea floor and the water in the immediate vicinity of cables.

In addition to the direct effects on fish due to construction and operation, indirect effects on fish are also conceivable. For instance, introduction of artificial hard substrata for the foundations on the seabed, which, in the North and Baltic Seas, is predominantly covered with fine sediments, could result in a change of the natural species composition.

Since trawl net fishing with beam trawls and trawl-door nets is banned within the wind park area, the pressure of fishing on the fauna would be reduced. Conceivably, such an area could be used as a refuge or spawning area.

The effects of the construction, existence and operation of offshore wind-power turbines on the fish fauna thus are largely temporary and spatially restricted to the area of the facility itself.

Since there are to date only very few data from investigations into the effects of offshore wind parks on the fish fauna, it is yet hardly possible to provide a scientifically sound answer to the question as to the effects on fish fauna in the context of concrete authorisation proceedings for offshore wind parks. However, as long as there is no adequate exact knowledge about the possible effects on the fish fauna, the assumption in the authorisation proceedings will be that there is no endangerment to fish, either during the construction phase or during the operational phase. Researches must therefore address both the effect of sound on fish (determination of hearing capacity, sound-determined behaviour reactions), and the possible reactions of fish to biotope changes during the construction and operational phases of the plants.

11 Distribution and Assemblages of Fish Species in the German Waters of North and Baltic Seas and Potential Impact of Wind Parks[1]

Siegfried Ehrich, Matthias HF Kloppmann, Anne F Sell, Uwe Böttcher

11.1 Introduction

The installation of wind parks could, through local alteration of habitat structures, potentially affect fish populations present in the area. To provide the most rigid analysis of this effect, the specific sites for planned wind parks should ideally be investigated through a multi-annual base line study before and another multi-annual study after installation of the turbines. However, this would be very cost-intensive, and we are not aware of any wind park where the entire fish community has been investigated to such an extend. With the following summary of ongoing independent long-term fisheries research within the two German Exclusive Economic Zones (EEZs) and the 12 nm territorial zones (German waters), we intend to provide the information currently available for predictions of the possible impact of new facilities.

Our interpretation of observations from these fisheries surveys is based on a number of fundamental characteristics of the fish and their habitats in the German waters of the North and Baltic Seas: A fish species will occur in a certain maritime area either if prerequisites exist which allow the species to stay in this particular suitable habitat for an extended time period, or if the species passes through the area while migrating to another area (e.g. feeding or spawning migrations). Alternatively, adverse circumstances may displace a fish species to a region it does not otherwise inhabit. The typical fish fauna found associated with a habitat belongs to the first category. Good examples of the second category are anadromous fish species that migrate from rivers into the sea for feeding and return to the rivers for spawning. The third category includes "Irrgäste", species that

---

[1] The research on distribution and assemblages of fish species in the German North and Baltic Seas was no part of the accompanying ecological research of the Federal Ministry for the Environment, Nature Conservation and Nuclear Safety.

enter a more or less hostile area as a result of bad weather conditions and an inflow of water masses from other areas.

To what habitat characteristics do fish respond? The availability of food and hydrographical parameters like water temperature and salinity are important for the survival and reproduction of most fish species. Other important habitat properties include water depth, chemical properties such as sufficiently high oxygen concentrations in the sea water (especially in the Baltic Sea, where in late summer the deep waters are often oxygen-depleted) and, particularly for the near-bottom fish fauna, the type of sediment of the sea ground.

The German North Sea is an area in which habitat characteristics can change not only seasonally but also over a period of days. To be successful, fish species must adapt to this variable habitat. Heavy rain expands the fresh water plume from the rivers into the German Bight, decreases the salinity of the near-shore water masses and shifts the river plume fronts further into the sea. The seasonal differences in bottom water temperature of the German North Sea (variation up to 15°C) are the highest in the entire North Sea. Cold winters with ice cover in the river mouths and in the Wadden Sea cause bottom water temperatures in the inner German Bight to drop to less than 5°C, which kills most stationary benthic animals down to a water depth of 35 m and leads the fish to migrate to deeper zones. On the other hand, in most parts of the EEZ the water column is, due to strong tidal currents, well-mixed even during hot summers, with bottom temperatures reaching up to nearly 20°C. In the mostly shallow waters, the turbulence of the waves during heavy gales reaches the bottom and whirls up the sediment. The high turbidity affects not only the benthic infauna and epifauna, but also the vertical distribution of fishes (Ehrich and Stransky 1999) and their dominating reproductive strategy, with pelagic eggs that cannot be buried by sediment.

Unlike the North Sea, the Baltic Sea lacks strong tidal currents, and its waters are also much less saline. Due to its strong fresh water inflow, predominantly in the Northeast, low evaporation rates, and its limited connection to North Sea waters, surface salinity values decrease from West to East. The German Baltic Sea areas belong to the transition zone between the North Sea and the Baltic Proper. Surface salinities range from > 17 psu in the Kiel Bight to < 8 psu in the Pomeranian Bight. In the bottom layers higher salinity values can be obtained as a consequence of highly variable intrusion of North Sea waters through the Danish straits. Consequently, the fish fauna of the Baltic is a mixture of relatively euryhaline marine and fresh water species, with a decreasing number of species towards the east (Hempel and Nellen 1974). Particularly in summer the gas exchange between waters in the bottom layers and surface waters is very restricted by a strong pycnocline. Thus, oxygen depletion in the bottom layers is a common feature of the deeper Baltic Sea areas in late summer.

Long-term investigations of the fish fauna in both, the German North Sea and Baltic Sea areas have been conducted by the German Federal Research Centre for Fisheries (BFA) throughout the last 50 years in the North Sea and the past 25 years in the Baltic Sea. We are using these long-term observations coupled with the results of an extended survey in 2004 to describe changes in occurrence and distribution of the fish species along different time scales and to consider potential sources of effect.

11.2   Material and Methods

The species lists, the frequencies of occurrence, the abundance of individual fish species, and the analyses of the bottom fish assemblages are based on data given in Table 1.

Table 1. Fisheries surveys included in this analysis. For investigations during each particular year or time period, the quarter, the area, the vessels, gears, number of hauls and the analyzed parameters are listed. The "entire area" refers to either German North or Baltic Sea. Abbreviations: FRV = Fisheries Research Vessel, WH = Walther Herwig

	Period/year	Quarter	Area	Vessels	Gears	Hauls	Parameters analyzed
North Sea	1958-2005	all	entire area	all FRV's	All	6,791	frequency of occurrence
	2004	4	entire area	Solea	cod hopper	57	frequency; abundance; assemblages
	1987-2005	all	Box A	WH II + III	GOV	1,007	frequency; assemblages
	2004	3	Box A	WH III	GOV	27	frequency of occurrence
Baltic Sea	1977-2005	all	entire area	all FRV's	All	2,045	frequency of occurence
	1991-2002	1	entire area	Solea	diff. bot. trawls	136	assemblages
	1991-2001	4	entire area	Solea	diff. bot. trawls	489	frequency; assemblages

During the period from 1958 onwards, catch data were sampled aboard six German fisheries research vessels, which used a large variety of gears, beam-trawls and otter trawls, equipped with various ground ropes. A description of the standard gear (GOV- chalut au Grand Ouverture Vertical) recently in use aboard 'Walther Herwig III' can be found in the manual of the International Bottom Trawl Survey (IBTS) (ICES 1999). The standard cod hopper aboard 'Solea' is comparable to the GOV as tested by comparison fishing experiments between both vessels.

In autumn 2004, a new survey (German Autumn Survey EEZ-German Bight – GASEEZ) aboard 'Solea' was initiated to describe the fish fauna in the German North Sea waters in greater detail, and to monitor the recent changes observed in that area due to climate warming (Corten and van de Kamp 1996; Ehrich and Stransky 2001). Predefined fixed stations were selected to cover specific depth ranges and to link to permanent stations for benthos monitoring conducted by the "Research Institution Senckenberg" (FIS) and "Alfred Wegener Institute of Polar and Marine Research" (AWI). Sampling stations covered eight spatial units in the German North Sea, which were defined as "benthos areas" by Rachor and Nehmer (2003) based on the distribution of benthos communities. Here, the "benthos areas" are considered as regions potentially relevant to the distribution of fish, not only due to the species composition of the macro-zoobenthos but also due to differences in the characteristics of sea bed topography, sediment type and the hydrography of the overlaying water column (Fig. 1 and Table 2). In November of 2004, fishing hauls were taken at 57 fixed stations (Fig. 1).

Table 2. Number of sampled stations per benthos area and depth range

	Benthos area	No. of stations	Depth range [m]
1	*Macoma balthica* community	4	15-21
2	*Goniadella-Spisula* II comm.	6	16-25
3	*Goniadella-Spisula* I comm.	5	20-30
4	*Nucula nitidosa* comm.	9	25-41
5	*Tellina fabula* comm.	10	20-44
6	*Amphiura filliformis* comm.	10	42-50
7	*Bathyporeia-Tellina* comm.	9	32-52
8	*Myriochele* Central North Sea comm.	4	44-66

Potential areas selected for the construction of offshore wind parks and for sand and gravel extraction cover a spatial range that is well-represented in the meso-scale dimension of the standard monitoring areas of the BFA survey (so-called "Boxes", 10-by-10 nm areas; Ehrich et al. 1998). Catches from Box A are comprised of 1,007 standard hauls with a standard otter trawl (GOV), which were carried out since 1987 (Fig. 1 and Table 1). These data constitute the basis to illustrate the changes in typical fish assemblages and abundance indices in single fish stocks of the German EEZ over time.

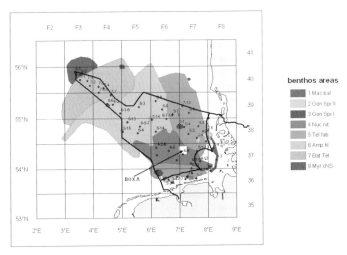

Fig. 1. Macro-benthos communities (benthos areas; details see Table 2) and locations of fishing stations (black dots). Solid black line: border of German North Sea waters. Location of Box A: 54°17' - 54°27' N, and 06°58' - 07°15' E

For the Baltic, a long-term survey of comparable meso-scale resolution does not exist. Still, on a larger spatial scale high and regular sampling effort has been maintained in the Lübeck Bight as well as in the Arkona Sea (Fig. 2). Complete annual surveys exist for both areas for the 4th quarter since 1991 until 2002.

In order to illustrate the variability of fish assemblages in the German North and Baltic Sea waters standardised fisheries data from both seas are compared. In the North Sea, data from methodologically consistent sampling between 1982 and 2002, in the Baltic Sea between 1991 and 2002 were selected (Table 1). The stations included in this analysis were selected to cover the regions of the potential offshore areas for wind farms and marine protected areas proposed by the Federal Agency for Nature Protection (Bundesamt für Naturschutz – BfN). However, because some of the areas were represented by very few or no historical fishery stations, the areas of analysis had to be extended (Figs. 3 and 4). The potential marine protected area 'Borkum Riffgrund' and the prospective wind farm area 'Borkum' were combined to one area 'Borkum'.

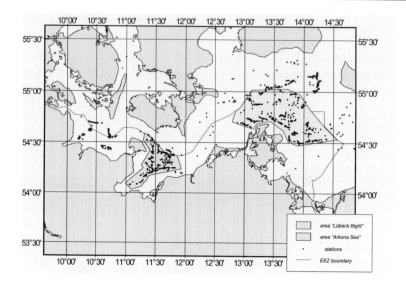

Fig. 2. Areas selected for trend estimation in composition and abundance of fish stocks in the Baltic Sea

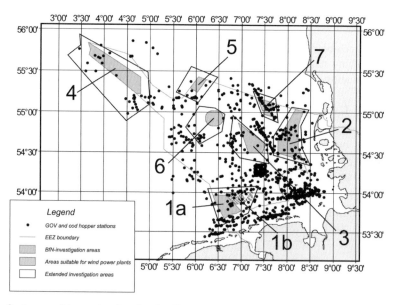

Fig. 3. Areas of investigation in the German waters of the North Sea. 1a: Borkum Riffgrund, 2: Amrum-Außengrund, 3: Osthang Elbe-Urstromtal, 4: Doggerbank, 5: Trittstein Elbe-Nord Urstromtal, 6: Trittstein Elbe-Mitte Urstromtal, 7 and 1b: potential wind power plant areas Sylt (7) and Borkum (1b)

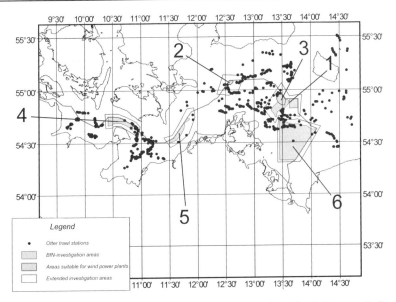

Fig. 4. Areas of investigation in the German Baltic Sea. 1: Adlergrund, 2: EG Kriegersflak, 3: EG Westlich Adlergrund, 4: Fehmarn-Belt, 5: Darsser Schwelle/ Kadettrinne, 6: Oderbank

The area swept by the gear was calculated using the wing-end spread and the towed distance. The abundance of species was then standardised to 1 ha (10,000 m²). For describing the bottom fish assemblages, the pelagic species were excluded from the analyses. The relatively high catches of pelagic species would otherwise disguise the real relationships within a bottom fish assemblage by single disproportionately high catches of herring, sprat or horse mackerel. The PRIMER-v5 software package (Clarke and Warwick 1994) was used to analyse the catch data for similarity of species composition. Multidimensional Scaling (MDS) based on the Bray-Curtis index based on square root or presence and absence transformed data was utilised in order to detect significant differences in fish assemblages between the defined areas. Similarity percentages were applied to identify the characteristic species. A preliminary analysis showed that separating the North Sea data into three different time intervals does not isolate differences in the assemblages of fish between the intervals. We therefore present one interval as a representative example.

11.3   Results for the North Sea - Assemblages at Different Spatial and Temporal Scales

11.3.1 German North Sea Waters: 1958 - 2005

Within the German North Sea waters, 102 fish species were collected in 6,791 hauls over the last 50 years by different vessels and gears (data from various otter trawls as well as beam trawls, and pelagic nets were combined, Table 3). The five species most frequently present in the hauls were the flatfishes dab (*Limanda limanda*) and plaice (*Pleuronectes platessa*), the gadoids whiting (*Merlangius merlangus*) and cod (*Gadus morhua*), and the pelagic clupeid herring (*Clupea harengus*), all of which were present in more than 50 % of the hauls.

Table 3. List of species in the German North Sea and Baltic waters, and sub-areas. Frequency of species' occurrence in hauls, number of hauls, and period of investigation are given

species name	common name	North Sea				Baltic Sea
		German waters	Box A	Box A		German waters
		frequency of occurrence in total number (n) of hauls, [%]				
		n=6,791	n=57	n=1,007	n=27	n=2,045
		1958-2005	Nov. 2004	1987-2005	August 2004	1977-2005
Agonus cataphractus	Hooknose	44.56	24.56	27.90	18.52	2.74
Alosa alosa	Allis shad	0.01	-	-	-	-
Alosa fallax	Twaite shad	9.12	40.35	11.72	48.15	0.34
Ammodytes marinus	Lesser sandeel	3.42	5.26	3.08	-	-
Ammodytes tobianus	Smooth sandeel, Lesser sandeel	0.15	-	-	-	2.49
Anarhichas lupus	Atlantic wolffish	0.19	-	-	-	-
Anguilla anguilla	European eel	1.19	1.75	-	-	9.14
Aphia minuta	Transparent goby	0.16	-	-	-	0.44
Argentina sphyraena	Argentine	0.03	-	0.10	-	-
Arnoglossus laterna	Scaldfish	16.86	12.28	26.32	96.30	0.20
Aspitrigla cuculus	Red gurnard	0.04	-	-	-	-
Atherina presbyter	Sand smelt	0.13	-	-	-	-
Belone belone	Garpike, garfish	0.24	-	0.20	-	-
Buglossidium luteum	Solenette	23.84	15.79	47.77	100.00	-
Callionymus lyra	Dragonet	36.36	47.37	64.75	100.00	0.15
Callionymus maculates	Spotted dragonet	3.37	8.77	0.89	-	-

Table 3. (cont.)

species name	common name	North Sea				Baltic Sea
		German waters	Box A	Box A		German waters
		frequency of occurrence in total number (n) of hauls, [%]				
		n=6,791 1958-2005	n=57 Nov. 2004	n=1,007 1987-2005	n=27 August 2004	n=2,045 1977-2005
Callionymus reticulatus	Reticulated dragonet	2.41	7.02	2.38	-	-
Chelon labrosus	Thicklip grey mullet	0.01	-	-	-	-
Chimaera monstrosa	Rabbit fish	0.01	-	-	-	-
Ciliata mustela	Five-bearded rockling	5.43	8.77	1.19	-	-
Clupea harengus	Herring	51.72	100.00	77.26	85.19	83.37
Coregonus lavaretus	Houting	-	-	-	-	0.05
Crenilabrus melops	Corkwing	-	-	-	-	0.10
Crystallogobius linearis	Crystal goby	0.16	-	-	-	0.20
Ctenolabrus rupestris	Goldsinny wrasse	0.06	-	-	-	-
Cyclopterus lumpus	Lumpsucker	2.47	3.51	0.50	-	5.77
Dicentrarchus labrax	European seabass	0.15	-	0.20	-	-
Echiichthys vipera	Lesser weever	3.03	21.05	1.39	11.11	-
Engraulis encrasicolus	European anchovy	2.89	29.82	6.55	14.81	2.84
Entelurus aequoreus	Snake pipefish	0.15	3.51	0.20	3.70	0.05
Esox lucius	Pike	-	-	-	-	0.05
Eutrigla gurnardus	Grey gurnard	42.85	68.42	80.44	100.00	1.17
Gadus morhua	Cod	59.96	87.72	75.97	22.22	97.46
Gaidropsarus vulgaris	Three-bearded rockling	0.56	-	0.40	-	-
Galeorhinus galeus	Tope shark	0.28	-	0.30	-	-
Gasterosteus aculeatus	Three-spined stickleback	6.45	-	2.58	-	0.29
Gobius niger	Black goby	-	-	-	-	0.54
Glyptocephalus cynoglossus	Witch	0.28	-	0.30	-	-
Helicolenus dactylopterus	Blue-mouth	0.12	-	0.40	-	-
Hippoglossoides platessoides	Long rough dab	13.12	36.84	7.35	-	1.91
Hippoglossus hippoglossus	Atlantic halibut	0.16	1.75	-	-	-
Hyperoplus immaculatus	Greater Sandeel	0.18	-	-	-	-
Hyperoplus lanceolatus	Greater Sandeel	8.28	8.77	4.67	14.81	1.42
Lampetra fluviatilis	River lamprey	1.56	14.04	0.99	11.11	0.20
Lepidorhombus whiffiagonis	Megrim	0.18	-	0.30	-	-
Limanda limanda	Dab	87.22	100.00	99.40	100.00	75.70
Liparis liparis	Striped sea snail	2.30	19.30	0.30	-	0.05
Liparis montagui	Montagu's sea snail	0.34	-	-	-	-
Lophius piscatorius	Angler	0.27	3.51	0.10	-	-

Table 3. (cont.)

species name	common name	North Sea				Baltic Sea
		German waters	Box A	Box A		German waters
		frequency of occurrence in total number (n) of hauls, [%]				
		n=6,791 1958-2005	n=57 Nov. 2004	n=1,007 1987-2005	n=27 August 2004	n=2,045 1977-2005
Lota lota	Burbot	-	-	-	-	0.15
Lumpenus lampretaeformis	Snake blenny	0.03	-	-	-	0.54
Maurolicus muelleri	Pearlsides	0.09	-	0.10	-	-
Melanogrammus aeglefinus	Haddock	6.86	8.77	8.14	-	3.42
Merlangius merlangus	Whiting	69.09	96.49	97.52	100.00	64.21
Merluccius merluccius	European hake	4.23	-	11.82	-	0.39
Micrenophrys lilljeborgi	Norway bullhead	0.04	-	-	-	-
Micromesistius poutassou	Blue whiting	0.03	-	-	-	-
Microstomus kitt	Lemon sole	21.62	36.84	42.50	-	0.68
Molva molva	Ling	0.19	-	-	-	-
Mullus surmuletus	Red mullet	6.05	35.09	19.86	18.52	0.98
Mustelus asterias	Starry smooth-hound	0.10	-	0.10	-	-
Mustelus mustelus	Smooth-hound	0.04	-	-	-	-
Myoxocephalus scorpius	Bull-rout, Short-spined sea Scorpion	31.53	40.35	20.95	22.22	11.79
Nerophis ophidion	Straight-nosed Pipefish	0.01	-	-	-	-
Oncorhynchus mykiss	Rainbow trout	-	-	-	-	0.24
Osmerus eperlanus	European smelt	4.77	14.04	0.10	-	10.02
Perca fluviatilis	Perch	-	-	-	-	0.10
Petromyzon marinus	Sea lamprey	0.12	-	0.50	-	0.05
Pholis gunnellus	Butterfish, Rock gunnel	0.59	-	-	-	-
Phrynorhombus norvegicus	Norwegian topknot	0.68	-	0.50	-	-
Platichthys flesus	European flounder	35.66	35.09	18.17	3.70	88.07
Pleuronectes platessa	Plaice	83.11	89.47	93.55	100.00	69.10
Pollachius pollachius	Pollack	0.28	-	0.10	-	0.54
Pollachius virens	Saithe	0.56	-	0.89	-	1.17
Pomatoschistus microps	Common goby, Sand goby	0.56	-	-	-	-
Pomatoschistus minutus	Common goby, Sand goby	8.20	56.14	21.55	51.85	0.73
Pomatoschistus pictus	Painted goby	0.01	-	-	-	0.05
Psetta maxima	Turbot	21.07	15.79	17.68	44.44	60.00
Raja batis	Blue skate	0.10	-	-	-	-
Raja clavata	Thornback ray	0.09	-	-	-	-
Raja radiata	Starry ray	1.46	7.02	-	-	-
Raniceps raninus	Tadpole-fish	0.24	-	-	-	-

Table 3. (cont.)

species name	common name	North Sea				Baltic Sea
		German waters		Box A	Box A	German waters
		frequency of occurrence in total number (n) of hauls, [%]				
		n=6,791 1958-2005	n=57 Nov. 2004	n=1,007 1987-2005	n=27 August 2004	n=2,045 1977-2005
Rhinonemus cimbrius	Four-bearded rockling	6.92	28.07	14.90	33.33	11.25
Rutilus rutilus	Roach	-	-	-	-	0.98
Salmo salar	Atlantic salmon	0.10	-	0.20	-	1.42
Salmo trutta	Sea trout, Brown trout	0.21	-	0.30	-	1.47
Sardina pilchardus	European pilchard, Sardine	3.39	22.81	14.80	14.81	-
Scomber scombrus	Atlantic mackerel	18.17	10.53	63.06	100.00	3.13
Scomberesox saurus	Altlantic saury	0.01	-	-	-	-
Scophthalmus rhombus	Brill	12.12	5.26	6.95	18.52	3.62
Scyliorhinus canicula	Small-spotted catshark	0.32	3.51	0.60	-	-
Scyliorhinus stellaris	Nursehound, Large-spotted dogfish	0.01	-	0.10	3.70	-
Sebastes viviparus	Norway redfish, Lesser redfish	0.03	-	0.20	-	-
Solea lascaris	Sand sole	0.01	-	-	-	-
Solea vulgaris	Common sole	38.42	3.51	11.92	3.70	1.86
Spinachia spinachia	Fifteen-spined stickleback, Sea stickleback	0.07	-	-	-	0.15
Sprattus sprattus	Sprat	49.29	96.49	71.10	77.78	72.32
Squalus acanthias	Spotted spiny dogfish, Spur-dog	0.65	3.51	0.10	-	-
Stizostedion lucioperca	Perch pike	-	-	-	-	2.54
Syngnathus acus	Greater pipefish, Common pipefish	0.10	-	-	-	0.15
Syngnathus rostellatus	Lesser pipefish	8.36	33.33	8.34	7.41	0.39
Syngnathus typhle	Snouted pipefish	0.18	-	0.40	-	0.24
Taurulus bubalis	Sea scorpion	0.09	-	0.10	-	-
Trachinus draco	Greater weever	0.09	-	-	-	0.64
Trachurus trachurus	Atlantic horse mackerel	26.74	68.42	81.53	100.00	22.44
Trigla lucerna	Tub gurnard	14.86	10.53	29.79	92.59	0.39
Trisopterus esmarki	Norway pout	1.00	-	0.79	-	0.34
Trisopterus luscus	Bib, Pouting	9.44	3.51	12.02	11.11	0.15
Trisopterus minutus	Poor cod	10.06	17.54	19.07	11.11	0.34
Zeus faber	John dory	0.06	7.02	0.10	3.70	-
Zoarces viviparus	Eelpout, Viviparous blenny	2.86	1.75	-	-	3.42
total number of species		**102**	**50**	**69**	**35**	**64**

11.3.2 German North Sea: Year of 2004

In the 57 catches of the survey in 2004, covering the German North Sea, 50 species could be detected, of which herring and dab were caught in 100 % of the hauls, followed by whiting, sprat and plaice, with frequencies of around 90 % and more (Table 3). Surprisingly, twaite shad *(Alosa fallax)* an Annex II and V species of the Species and Habitat Directive, which had occurred only as single specimens during former decades, and also the lesser weaver *(Echiichthys vipera)*, a red list species (Fricke et al. 1995), were caught very frequently in 2004 (40 % and 21 % of all hauls, respectively).

In terms of catch in numbers of the 2004 survey, the six most abundant species sprat, herring, sand goby, dab, grey gurnard and horse mackerel, accounted for more than 95 % of the total catch, not only in the German waters as a whole, but also in the individual benthos areas (Table 4). Based on catch in weight the dominant species was dab, followed by grey gurnard, sprat, herring, plaice and whiting. Together, they contributed more than 90 % of total catch in weight. The species compositions of the eight benthos areas are also listed. For the reasons given above the pelagic species were excluded from the analysis when comparing the bottom fish assemblages. Therefore the two most abundant species in five of the eight areas (sprat and herring) had no influence on the results of the analysis of similarity presented in a MDS-plot (Fig. 5).

As expected, the bottom fish assemblages in the near-shore benthos areas were very similar. But the dissimilarities increased with increasing depth and distance from the coast, lower bottom temperatures, the existence of a thermocline and higher salinities. The fish assemblages of the benthos areas 1 to 5 largely overlapped and were barely separable. In spite of relatively shallow waters at the tail-end of the 'Doggerbank', benthos area 7 was clearly different from the more coastal areas with similar water depths. Benthos area 6 was inhabited by a fish assemblage which formed a link between the near-shore (3 - 5) and the offshore areas (7 and 8). There was no overlap of the community in benthos area 8 with any of the other areas, the fish assemblage of this deepest area within the German waters included elements of the more northern and Atlantic-influenced assemblages.

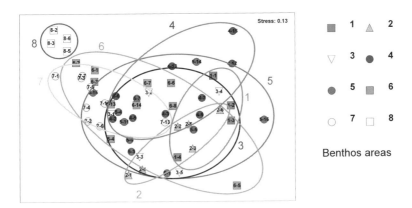

Fig. 5. MDS ordination of the stations based on the species similarity matrix. The distance between two data points corresponds to the amount of dissimilarity between the two corresponding species compositions. The ellipses are drawn by hand to include all samples from the respective benthos area (see Fig. 1 and Table 2)

The species characteristic (up to 90 % similarity percentages) for benthos areas 1 - 8 are marked in Table 4. The most important of these 14 characteristic species were dab, whiting and sandy goby, followed by grey gurnard and plaice. Rare species of mean abundance of less than 0.5 per ha (10,000 m²) were without any significance for the results of this analysis.

Table 4. List of species in the German North Sea and sub-areas present in the November 2004 survey. Depth range sampled and numbers of hauls are given. Values for species characterising an assemblage are marked by grey fields.

	entire area	benthos area							
		1	2	3	4	5	6	7	8
	15-66m	15-21m	16-25m	20-30m	25-41m	20-44m	42-50m	32-52m	44-66m
	57 hauls	4 hauls	6 hauls	5 hauls	9 hauls	10 hauls	10 hauls	9 hauls	4 hauls
species	mean abundance [individuals per ha]								
Sprattus sprattus	1295.11	733.69	520.30	1548.00	507.00	4546.00	908.30	92.09	22.52
Clupea harengus	332.66	72.32	83.10	936.80	424.33	429.20	410.70	64.75	172.31
Pomatoschistus minutus	301.03	68.95	16.84	212.80	1469.00	197.30	50.80	1.75	-
Limanda limanda	185.47	12.73	11.39	101.20	210.78	204.90	427.40	39.27	343.33

Table 4. (cont.)

species	entire area	benthos area							
		1	2	3	4	5	6	7	8
	15-66m	15-21m	16-25m	20-30m	25-41m	20-44m	42-50m	32-52m	44-66m
	57 hauls	4 hauls	6 hauls	5 hauls	9 hauls	10 hauls	10 hauls	9 hauls	4 hauls
		mean abundance [individuals per ha]							
Eutrigla gurnardus	30.97	-	0.05	13.80	1.11	7.80	58.70	48.82	145.34
Trachurus trachurus	26.35	-	0.02	3.80	18.78	24.30	90.70	10.21	17.99
Pleuronectes platessa	18.33	0.28	0.21	6.80	32.11	29.30	39.50	2.90	1.37
Merlangius merlangus	16.45	0.53	4.49	44.00	17.00	20.00	24.00	6.33	9.61
Engraulis encrasicolus	4.88	-	-	0.60	0.56	6.90	13.00	7.53	0.81
Gadus morhua	2.18	0.25	1.10	4.40	2.89	3.50	2.20	0.53	1.68
Alosa fallax	1.93	0.53	0.28	1.80	9.44	1.20	-	-	-
Hippoglossoides platessoides	1.92	-	-	0.40	0.44	0.30	8.60	0.26	2.95
Sardina Pilchardus	1.60	-	-	-	0.11	1.00	2.50	5.83	0.67
Echiichthys Vipera	1.59	-	0.12	15.40	0.11	0.60	0.10	0.55	-
Platichthys flesus	1.26	0.42	0.16	0.80	4.56	2.40	-	-	-
Microstomus kitt	1.22	-	-	-	0.89	0.10	2.10	0.61	8.56
Callionymus lyra	1.18	0.04	0.07	1.00	3.11	0.60	2.50	0.25	0.14
Melanogrammus aeglefinus	0.96	-	-	-	-	-	-	0.36	12.91
Syngnathus rostellatus	0.88	0.28	1.15	2.00	2.22	1.20	-	-	-
Mullus Surmuletus	0.87	-	-	2.20	0.67	1.60	1.60	0.06	0.07
Agonus cataphractus	0.83	1.33	0.12	1.60	1.00	2.30	0.10	-	-
Rhinonemus cimbrius	0.81	-	0.05	0.40	2.56	-	2.10	0.02	-
Myoxocephalus scorpius	0.81	0.53	0.35	0.60	1.44	2.50	0.10	-	-
Liparis liparis	0.73	0.60	0.05	0.80	3.44	0.40	-	-	-
Buglossidium luteum	0.42	-	-	-	1.00	0.30	1.20	-	-
Trisopterus minutus	0.29	-	0.02	0.20	0.11	-	1.40	0.02	0.11
Osmerus eperlanus	0.24	0.14	-	1.00	0.67	0.20	-	-	-
Arnoglossus laterna	0.24	-	-	0.20	0.11	-	1.10	-	0.11
Psetta maxima	0.20	-	-	-	0.22	0.40	0.50	0.02	-
Lampetra fluviatilis	0.17	0.25	0.09	0.40	0.11	-	0.50	-	-
Trigla lucerna	0.13	-	-	-	0.33	-	0.40	0.03	-
Callionymus reticulatus	0.11	-	-	-	-	0.60	-	-	-
Ciliata mustela	0.08	0.18	0.02	0.20	-	0.30	-	-	-

Table 4. (cont.)

species	Entire area 15-66m 57 hauls	benthos area 1 15-21m 4 hauls	2 16-25m 6 hauls	3 20-30m 5 hauls	4 25-41m 9 hauls	5 20-44m 10 hauls	6 42-50m 10 hauls	7 32-52m 9 hauls	8 44-66m 4 hauls
					mean abundance [individuals per ha]				
Scomber scombrus	0.07	-	-	-	-	0.20	0.10	0.08	0.04
Callionymus maculatus	0.06	-	0.02	-	0.22	0.10	-	-	0.07
Hyperoplus lanceolatus	0.04	0.04	0.02	0.20	-	0.10	-	0.02	-
Ammodytes marinus	0.04	-	0.02	-	-	0.20	-	-	-
Scophthalmus rhombus	0.04	-	-	-	-	0.10	0.10	0.02	-
Solea vulgaris	0.04	-	-	-	0.22	-	-	-	-
Cyclopterus lumpus	0.02	-	-	-	0.11	-	-	-	0.04
Entelurus aequoreus	0.02	-	-	0.20	-	-	-	0.02	-
Scyliorhinus canicula	0.02	-	-	0.20	-	-	-	-	0.04
Squalus acanthias	0.02	-	-	-	-	-	0.10	0.02	-
Trisopterus luscus	0.02	-	0.02	-	-	-	0.10	-	-
Anguilla anguilla	0.02	-	-	-	-	-	0.10	-	-
Zoarces viviparous	0.02	-	-	-	-	0.10	-	-	-
Zeus faber	0.02	-	-	-	-	-	-	0.08	0.07
Raja radiata	0.01	-	-	-	-	-	-	0.05	0.11
Lophius piscatorius	0.00	-	-	-	-	-	-	-	0.07
Hippoglossus hippoglossus	0.00	-	-	-	-	-	-	-	0.04
total number of species	**50**	**18**	**25**	**29**	**33**	**33**	**30**	**28**	**25**

11.3.3 Potential Wind Park Sites and Marine Protected Areas: 1982 - 2002

All in all, 63 species of fish have been recorded in those areas for the entire time interval 1982 - 2002 (Table 3). Pelagic clupeids, herring and sprat, appeared to be the most abundant species in almost all selected areas when judged by mean catch per hour. Of all demersal fish, dab, whiting and plaice were the most abundant species. In the eastern areas of the German North Sea dab catches reached values of more than 1,000 individuals per trawled hour, whereas in the western (offshore) areas they were much less abundant. The highest densities of whiting were recorded in some of the northern areas of the German EEZ while plaice appeared to be most abundant in the central German Bight (Table 5).

Table 5. Species caught in selected areas of the German North Sea between 1982 and 2002. Values in mean catch per 30 minutes [individuals/30 min]

species name	Amrum	Borkum	Dogger-bank	EG Sylt	Osthang Elbe-Urstrom	Trittstein Elbe Nord	Trittstein Elbe Mitte
Agonus cataphractus	5.78	0.41	0.04	0.73	0.59	-	0.29
Alosa fallax	0.08	0.06	-	-	1.10	-	-
Ammodytes sp.	0.28	0.05	-	-	0.14	-	-
Ammodytes marinus	1.02	0.13	-	0.05	0.17	-	-
Ammodytes tobianus	4.17	-	-	-	-	-	0.07
Aphia minuta	0.02	-	-	0.05	0.04	-	-
Arnoglossus laterna	0.04	-	0.11	0.06	1.12	0.25	0.22
Buglossidium luteum	0.35	0.42	0.32	0.41	29.12	-	0.65
Callionymus sp.	-	-	-	0.13	-	-	-
Callionymus lyra	0.57	0.19	1.15	0.62	1.42	4.67	2.15
Callionymus maculatus	0.91	0.03	-	0.09	0.40	-	-
Callionymus reticulans	-	-	-	-	0.04	-	-
Ciliata mustela	0.12	0.05	-	-	0.02	-	-
Clupea harengus	1,766.88	1,453.74	3,007.45	819.33	1,474.54	10,092.34	5,976.11
Dicentrarchus labrax	0.01	-	-	-	-	-	-
Echiichtys vipera	0.07	0.30	-	-	-	-	-
Engraulis encrasicolus	-	0.02	-	-	0.40	-	0.04
Entelurus aequoreus	-	-	-	-	0.04	-	-
Eutrigla gurnardus	0.08	0.05	43.34	0.45	1.20	8.13	0.79
Gadus morhua	6.54	9.39	6.43	2.96	8.10	9.04	3.40
Gaidropsarus vulgaris	0.03	-	-	0.03	0.20	-	-
Gasterosteus aculeatus	1.15	0.25	-	0.08	0.10	-	-
Glyptocephalus cynoglossus	-	0.01	0.04	-	-	0.04	-
Gobiidae gen. sp.	0.53	0.49	0.04	0.34	0.17	-	-
Hippoglossoides platessoides	-	0.03	13.36	0.14	10.09	26.13	15.47
Hyperoplus lanceolatus	0.93	0.14	0.47	0.21	0.05	-	-
Lampetra fluviatilis	-	0.01	-	-	-	-	-
Lepidorhombus whiff-iagonis	-	-	-	-	-	-	0.07
Limanda limanda	551.44	173.93	176.11	530.51	540.65	323.42	231.07
Liparis liparis	0.03	0.01	0.04	0.07	0.24	-	-
Lophius piscatorius	0.02	-	0.04	-	-	-	-
Maurolicus muelleri	0.02	-	-	0.01	0.04	-	-
Melanogrammus aeglefinus	-	0.02	4.97	-	4.44	4.46	0.50
Merlangius merlangus	160.05	42.20	121.88	25.89	123.62	431.59	159.75
Microstomus kitt	0.04	0.10	1.11	0.10	0.75	1.59	0.18
Mullus surmuletus	0.04	0.02	-	-	-	-	-
Myoxocephalus scorpius	4.70	0.36	-	0.48	0.72	-	0.07
Nerophis ophidion	-	-	-	-	-	0.09	-
Pholis gunnelus	0.03	0.02	-	-	0.04	-	-

Table 5. (cont.)

species name	Amrum	Borkum	Dogger-bank	EG Sylt	Osthang Elbe Urstrom	Trittstein	
						Elbe Nord	Elbe Mitte
Phrynorhombus norvegicus	-	-	-	0.03	-	-	-
Platichthys flesus	1.52	2.89	0.04	2.08	4.50	0.34	0.75
Pleuronectes platessa	18.89	29.32	8.27	18.03	177.77	22.79	23.22
Pollachius pollachius	-	-	-	0.03	-	-	-
Pollachius virens	0.02	0.01	-	-	0.04	-	-
Pomatoschistus minutus	3.49	0.20	-	3.50	2.50	-	-
Psetta maxima	0.07	0.05	0.04	-	0.15	0.21	0.11
Raja radiata	0.02	-	2.25	-	-	0.42	-
Raniceps raninus	0.01	-	-	-	-	-	-
Rhinonemus cimbrius	0.18	-	-	-	3.44	1.04	2.07
Sardina pilchardus	-	0.06	-	-	-	-	-
Scomber scombrus	-	-	-	-	0.07	-	0.07
Scophthalmus rhombus	0.04	0.03	0.06	0.03	0.04	-	0.07
Solea vulgaris	0.09	0.29	-	0.14	0.60	0.09	0.18
Sprattus sprattus	2,323.19	1,522.98	815.18	393.53	1,812.62	489.92	1,827.47
Squalus acantias	-	-	0.04	-	0.04	-	-
Syngnathus sp.	0.08	-	-	-	-	-	-
Synganthus rostellatus	1.55	0.14	0.04	0.51	0.17	-	-
Syngnathus acus	-	-	-	0.03	0.04	-	-
Trachinus draco	0.04	-	-	-	-	-	-
Trachurus trachurus	-	0.28	0.09	0.09	33.05	-	1.15
Trisopterus esmarki	-	-	8.32	-	-	-	0.29
Trisopterus luscus	0.15	0.23	0.04	0.12	0.10	0.17	0.22
Trisopterus minutus	0.35	0.33	0.90	0.12	0.79	14.29	1.36
Zoarces viviparus	0.04	-	-	-	-	-	-

Multidimensional Scaling (MDS) plots of species similarity matrices based on untransformed as well as square root-transformed data revealed no significant results. This is partly due to the relatively high variability of the data as a consequence of different gear and ship performance, but also underscores the relatively homogenous nature of the fish communities of the German waters. Differences between the selected areas only became apparent when the presence/absence transformation of the data was used. This transformation sets all abundance values to 1 if a species is present in the catch, and to 0 if it is absent, thus giving the rare species a higher impact in the structure of the distance matrix. Even with this transformation the similarity between all areas is quite high. However, it becomes apparent that the similarity between two arbitrarily selected stations decreases with distance from the coast (Fig. 6).

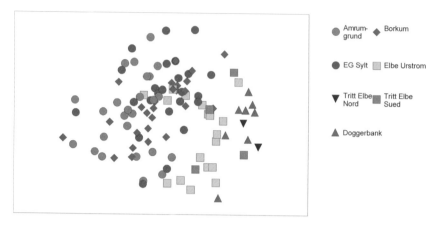

Fig. 6. MDS ordinations of the stations within the extended areas of investigation in the North Sea based on the species similarity matrix. The distance between two data points corresponds to the amount of dissimilarity between the two corresponding species assemblages (German North Sea areas 1991 - 2000)

11.3.4 Box A: 1987 - 2005

Among the five species most frequently caught in Box A were dab, plaice and whiting, as it was the case in the German North Sea waters in general, and additionally horse mackerel and grey gurnard – all with presence in more than 80 % of the hauls (Table 3). In total, 69 species were caught, a comparatively high number, which is in part the result of the high total number of hauls applied from 1987 - 2005. For instance, after three consecutive days of fishing in 2004 with a total of 27 hauls, the number of species caught was only 37.

For the analysis of temporal trends in bottom fish assemblages, catch data for the pelagic species were excluded, as well as all data from winter surveys (quarters 1 and 4). The 2-dimensional ordination of the species similarity matrix shows a conspicuous shift in fish assemblage from 1987 to 2005, particularly between the period until 1992 and the later years (1997 - 2005; Fig. 7).

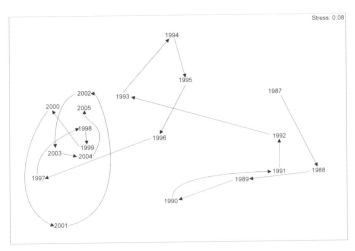

Fig. 7. MDS ordination of the mean species assemblages in Box A for the summer quarters of the years 1987 - 2005 based on the similarity matrix. The distance between two data points corresponds to the amount of dissimilarity between the two corresponding species assemblages

Comparison of abundance indices in mean catch per 30 min of selected demersal fish species suggests that the German Bight fish assemblage changed from a more or less gadoid (cod and whiting)-dominated (1987 - 1992) to a flatfish-dominated assemblage (1997 - 2005). The years between 1993 and 1996 constituted a transition, during which neither of these fish groups dominated. Also the increased abundance of smaller fish species as solenette (*Buglossidium luteum*) and scaldfish (*Arnoglossus laterna*) contribute conspicuously to the German Bight fish assemblage that has been typical in recent years. It is also noteworthy that southern species such as red gurnard, sardine, anchovy, and striped red mullet have appeared more regularly in the samples of recent years than during the early part of the investigation (Fig. 8; see also Ehrich and Stransky 2001).

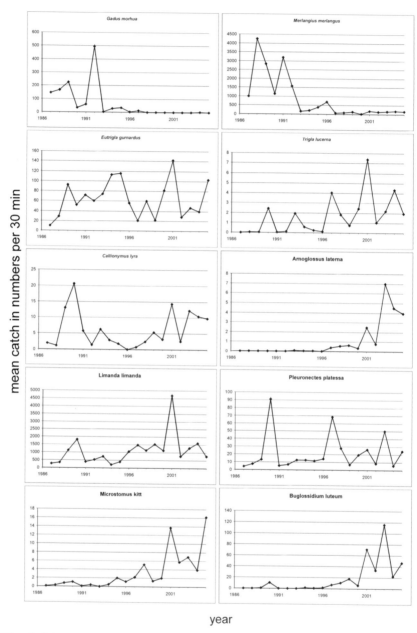

year

Fig. 8. Mean catch frequencies of selected fish species in Box A between 1987 and 2005 (note the differences in scale on the y-axes)

11.4   Results for the Baltic Sea - Assemblages at Different Spatial and Temporal Scales

11.4.1 Entire German Baltic Sea: 1977 - 2005

Within the German part of the Baltic Sea 63 fish species were collected in 2,045 hauls over the past 28 years by various vessels and gears (different otter trawls as well as pelagic nets; Table 3). Even though the number of fish species was smaller than in the German North Sea, more species occurred frequently in the catches. The nine species most frequently present in the hauls were the flatfishes dab (*Limanda limanda*), plaice (*Pleuronectes platessa*), flounder (*Platichthys flesus*), and turbot (*Psetta maxima*), the gadoids whiting (*Merlangius merlangus*) and cod (*Gadus morhua*), and the pelagic clupeids herring (*Clupea harengus*) and sprat (*Sprattus sprattus*), all of which were present in more than 50 % of the hauls. Though only rarely caught, the occurrence of fresh water fish like pike (*Esox lucius*), perch (*Perca fluviatilis*) and perch pike (*Stizostedion lucioperca*) in the catches appears noteworthy.

11.4.2 Potential Wind Park Sites and Marine Protected Areas: 1990 - 2002

All in all, 31 species were recorded during the first quarter in the selected areas of the German Baltic Sea. Most species occurred in the area next to Fehmarn (21 species) and around Kriegersflak (22 species). The lowest number of species (10) was recorded in the Darsser Schwelle area, where however only four hauls were taken during the ten years of investigation. Sixteen different species were recorded for each of the areas Adlergrund and Oderbank. On the Oderbank, fish abundance was by far the lowest (Table 6).

Again, sprat and herring were the most abundant species in almost all areas, however, with the same restrictions as mentioned for the North Sea. Of the demersal species, cod, dab and flounder were the most abundant. Cod was most abundant in the three central areas, while its abundance decreased to the west as well as to the east. Dab was most abundant in the west (Fehmarn), while its density decreased towards the east. The abundance of flounder increased towards the east, though its abundance at Oderbank was not as high as in the other eastern areas (Table 6).

Table 6. Species caught in selected areas of the German Baltic Sea between 1990 and 2002. Values in mean catch per 30 min [individuals/30 min]

Species	Fehmarn	Darsser Schwelle	Kriegersflak	Adlergrund	Oderbank
Agonus cataphractus	0.05	-	0.03	-	-
Ammodytes sp.	-	0.25	-	-	0.67
Anguilla anguilla	-	-	0.02	-	-
Clupea harengus	333.14	26.00	3863.81	5479.63	117.85
Cyclopterus lumpus	-	0.50	0.10	0.13	0.32
Eutrigla gurnardus	0.09	-	-	-	-
Gadus morhua	54.04	641.25	275.98	251.76	11.43
Hippoglossoides platessoides	1.46	-	-	-	-
Limanda limanda	123.49	20.00	9.05	0.95	1.36
Melanogrammus aeglefinus	-	-	0.03	-	-
Merlangius merlangius	84.19	0.50	33.60	4.03	-
Merluccius merluccius	-	-	0.03	-	-
Myoxocephalus scorpius	0,07	-	0.07	0.13	1.68
Osmerus eperlanus	-	-	0.13	41.98	0.36
Platichthys flesus	5.49	6.25	61.45	51.25	12.27
Pleuronectes platessa	0.87	0.75	5.07	6.60	3.93
Pollachius pollacius	0.05	-	0.28	0.05	-
Pollachius virens	0.05	-	0.07	-	-
Psetta maxima	0.48	2.50	1.10	1.22	2.51
Rhinonemus cimbrius	0.16	-	0.16	2.63	-
Rutilus rutilus	0.03	-	-	-	0.05
Salmo salar	-	-	0.10	0.05	0.08
Salmo trutta	0.05	-	0.10	-	-
Scomber scombrus	0.03	-	0.03	-	-
Scophthalmus rhombus	0.03	-	-	-	-
Solea vulgaris	0.03	-	-	-	-
Sprattus sprattus	615.14	23.25	3880.41	7079.35	79.05
Stizostedion lucioperca	-	-	-	-	0.03
Trisopterus minutus	0.05	-	-	-	-
Zoarces viviparus	-	-	0.23	0.03	0.02

As for the North Sea data, MDS plots based on Bray Curtis similarity after presence/absence transformation yielded the most significant results. There appears to be a strong overlap between all areas in species composition. However, similarity between stations generally decreased with increasing distance from west to east (Fig. 9).

In the detailed comparison of the easternmost and the westernmost areas, a total of 48 species were detected in the two areas selected for trend estimations of assemblage composition. Forty-two of these species occurred in the Arkona Sea and 36 in the area of Lübeck Bight. Only five species occurred solely in Lübeck Bight, while eleven species were found only in the area of Arkona Sea. The 2-dimensional ordination of the species similarity matrices suggests that a shift in the fish assemblage has taken place in both areas with the beginning of the 1990s. In all other years, the fish assemblages appeared to be quite similar – apart from the year 1997 in the Lübeck Bight. Data from before 1990 still need to be validated and have therefore not been taken into account. It is thus impossible to conclude whether there has been a real shift or whether the 1990 and, partly, the 1991 data represent outliers. However, the 1990 ordination corresponds very well with the low catches of cod in both areas, while the 1997 outlier corresponds with exceptionally high cod catches in the Lübeck Bight (Fig. 10).

Abundance developments of some selected species do not indicate conspicuous trends, which is in accordance with the MDS ordination. An exception might be the abundance of plaice in the Arkona Sea which shows an increasing trend, at least in this particular area. All other species are subject to more or less strong variability in abundance, but no directed trend was observed. Species such as eel (*Anguilla anguilla*), shorthorn sculpin (*Myoxocephalus scorpius*), fourbearded rockling (*Rhinonemus cimbrius*), and turbot (*Psetta maxima*) appear to have increased in abundance since 1990 and decreased again at the end of the 90s. The abundance of dab appears to follow a 5-year cycle of decrease and increase (Fig. 11).

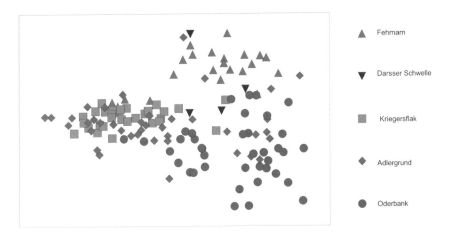

Fig. 9. MDS ordinations of the stations within the extended areas of investigation in the Baltic Sea based on the species similarity matrix. The distance between two data points corresponds to the amount of dissimilarity between the two corresponding species assemblages

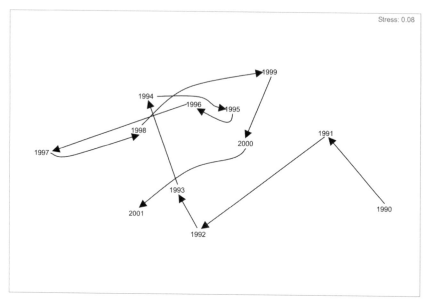

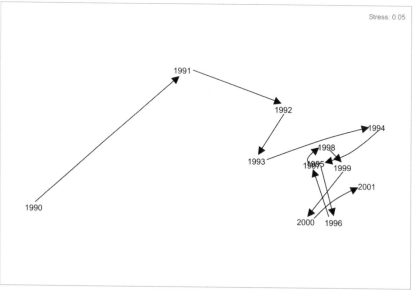

Fig. 10. MDS ordination of the mean fish assemblages based on the species similarity matrix for the period 1990 to 2001 in both areas, Lübeck Bight (top) and Arkona Sea (bottom). The distance between two data points corresponds to the amount of dissimilarity between the two corresponding species assemblages

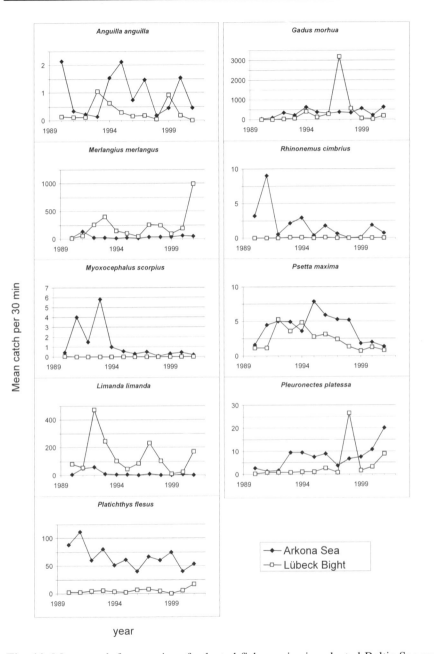

Fig. 11. Mean catch frequencies of selected fish species in selected Baltic Sea areas between 1990 and 2002

11.5   Discussion

The species lists for the German North Sea and Baltic Sea waters (Table 3) are based on cumulative analysis of hauls from a variety of fishing gears applied, from a beam trawl to a pelagic trawl without bottom contact. Beam trawls with tickler chains were used to catch flatfish buried in the sediment, such as sole. Pelagic trawls were applied to catch open-water species like herring and sprat. The gears used most often were otter trawls equipped with a rubber discs ground rope. These trawls like the GOV (chalut au Grand Ouverture Vertical) trawl, the ICES standard trawl for the North Sea, have only weak contact with the bottom, and the species living in or very close to the bottom, like solenette (*Buglossidium luteum*), hook-nose (*Agonus cataphractus*) and dragonets (Callionymidae) are quantitatively underrepresented in the catches. Other parameters of strong impact are the mesh size and especially the use of a fine-meshed liner inside the codend to prevent the escape of small fish. Therefore, the qualitative and quantitative species composition and the length distribution of fish in the hauls are not only area-specific but also gear-specific and might not be fully representative of the fish fauna living in that area. Therefore, when describing a fish assemblage by using data from a single gear, the caveat that the description is gear-dependent should be highlighted. In this paper the frequency of a species within the combined North Sea hauls can and should only give a semiquantitative representation of the species' occurrence.

In the 2004 survey, the six most abundant species were responsible for 95 % of the total catch in numbers, a value which is common for North Sea areas. Therefore the diversity is low compared to other, especially tropical or subtropical areas. The total number of fish species found in the German North Sea varies between 102 (Table 3) and 189 species, given by the authors of the Red List on marine fishes of the German Wadden Sea and North Sea (Fricke et al, 1995). Some of the 87 additional species in the latter list live in very shallow waters, others like big sharks are able to avoid the gears commonly used and yet others, like sun fish (*Mola mola*), stay at the surface, and thus outside the paths of the gears. Most of these species are vagrant species ("Irrgäste"), accidentally present and fished in an unsuitable habitat. Their rare attendance is of no importance to describe or assess the quality of the habitat. The bluemouth redfish (*Helicolenus dactylopterus*) is a good example. The adults normally live at the continental slope at depths from 400 to 800 m. This species is adapted to great depths and cold water. An invasion of very young blue-mouths in the northern North Sea (possibly in the course of an unusual strong inflow of Atlantic water) took place in the winter of 1990/91 (Heessen et al. 1996). Afterwards they have spread over a wide area, entering the German waters during the summer of 1992 (Box A) and were also fished in the Wadden Sea by German shrimpers during that time.

Analysis of long term fisheries data from both, German North Sea and Baltic Sea waters, have revealed no striking differences between selected areas within each sea, neither with untransformed nor with square root-transformed data. Only with presence/absence transformation has it been possible to group stations into significant clusters according to their geographical origin. Consequently, it is the occurrence of rare species which is able to characterise particular areas of the North and Baltic Seas. These species are in most cases small and more or less bound to their benthic habitat. Abundance differences in the dominant and in most cases larger and more mobile species are unable to achieve significant results in long term data analysis, because they occur almost everywhere.

In the German Bight it is mainly the depth and the distance from the coast that appears to have a major impact on the composition of the typical fish assemblage. Further separation into smaller faunal areas is almost impossible. Particularly in the coastal areas the environment is highly variable depending on the inflow of Atlantic waters from the south-west and the freshwater discharges of the major rivers. Changes in the wind field may alter the boundaries (fronts) of these water masses (Dippner 1993), where, due to accumulation of plankton, fish may also aggregate. This high variability in the hydrographic regime makes it impossible to define distinct fish assemblages in the near-coastal areas of the German Bight.

In the Baltic Sea, fish assemblages are influenced mainly by the salinity gradient, with salinities decreasing from west to east. The number of species also decreases towards the east, and only euryhaline marine and freshwater species that are able to cope with the brackish waters can be found there. Only during periodic intrusion of North Sea waters at depth may such less euryhaline species as the dab, the plaice or the whiting also be found in the eastern areas. Intermittent oxygen depletion due to the strong vertical stratification will, however, limit the occurrence of these fish in those areas, resulting in a relatively high variability in demersal fish assemblages over time.

11.5.1 Potential Impact of Wind Turbine Construction on Fish Assemblages

Introduction of hard substrate to sandy or muddy sea floor could, either in and of itself or by the ensuing colonisation by specialised benthic communities attract both demersal and pelagic fish. In addition to many anecdotal reports by fishermen and divers, scientific studies, too, have provided evidence of fish being attracted by underwater structures in northern latitudes (e.g. Løkkeberg et al. 2002).

Offshore Wind Farms

The first results on community effects of offshore wind farms have come almost exclusively from technical reports prepared for the few already existing wind farms in northern waters. So far, these observations are based on the developments of a few years, and hence cannot yet be used to describe or predict long-term trends. Still, they may be useful in indicating possible short-term responses, as they investigate the structures of greatest similarity to the new winds farms being proposed for the North and Baltic Seas.

Setting up the foundations of the Horns Rev offshore wind farm off Blåvands Huk (Denmark) introduced hard substrate within a sandy natural seabed in the northernmost part of Europe's Wadden Sea. The last of 80 turbines was installed in August 2002, and between 2002 and 2004, investigators found no indications that the construction of the wind farm had had any effect on the sediment composition in the surrounding area (the Wentworth sediment classes are silt/clay and very fine sand). This leads the authors to conclude that the wind farm is not likely to have any effect on sandeels in the area (Jensen et al. 2004). However, the observations are restricted to the short time span of two years.

An accompanying hydroacoustic survey of fish conducted at Horns Rev indicated only relatively weak short-term effects on the abundance of fish, with no significant difference between impact and reference areas (Hvidt et al. 2005). A tendency of increasing densities of large fish in the vicinity of a topographically varied seabed was observed, but this was seen both inside and outside the wind park area, and was based on a relatively small total number of large fish (n=184). Unfortunately, regular analyses of the assemblage of fish around Horns Rev were later discontinued.

Decommissioned North Sea Oil Platforms

Additional observations of the effects of artificially introduced hard substrate stem from decommissioned oil platforms where several studies indicated that fish are attracted by the structures. Fish densities in the vicinity were many times as high as at distances > 100 - 150 m away from the platforms (Løkkeborg et al. 2002). At Albuskjell, a platform in the central North Sea, the cod (*Gadus morhua*) was the most abundant species in the catches, while in the vicinity of Gullfaks in the northern North Sea, the ling (*Molva molva*) was the dominant species caught, followed by the saithe (*Pollachius virens*) and the cod. However, the authors warn that although their study provided evidence of pronounced aggregations of fish close to the platforms, responses are complex, and results are inconclusive regarding species-specific temporal and spatial patterns. E.g., at Gullfaks, only the ling was concentrated around the platform, while cod and saithe were evenly distributed. Possibly, interactions between these species lead cod and saithe to avoid areas with high densities of ling (Løkkeborg et al. 2002). Supporting evidence for this effect would need more extended observations (here included: for Gullfaks one, for Albuskjell two fishing cruises). In a study combining hydroacoustics and net hauls, Valdemarsen (1979) reported elevated fish densities within a 200 m radius around a North Sea oil platform (Ekofisk – installed on a flat seabed with muddy and sandy sediments), especially for saithe and to a lesser degree for cod. There are somewhat supporting (although not statistically verified) indications from hydroacoustic measurements and video observations at the Albuskjell platform that accompanied the study discussed above (Soldal et al. 2002).

Artificial Reefs

As an intended introduction of hard substrates, numerous artificial reefs – mostly in the tropics – have been built from dumped waste material or specific structures made for the purpose of increasing the densities of local fish populations (Svane and Petersen 2001). Near the Dutch coast at Noordwijk, basalt blocks in heaps of each about 1.5 m height and 12 m diameter have been placed on the sea floor. Fish and benthic fauna were assessed starting prior to the installation of the artificial reef in 1992 (although only through a single 4-day cruise) and afterwards until 1995 for benthos and until 1993 for fish (Leewis and Hallie 2000). The reef showed steadily increasing biomass and diversity of typical North Sea benthos over the time interval investigated. Unfortunately, in this case too, the investigations of effects on the fish assemblage were terminated before long-term effects could be observed. In the U.K., Torness artificial reef, built in 1984 off the south-eastern coast of Scotland from quarried rock, apparently influenced local populations of cod. A study conducted from 1988 to 1990 showed higher densities of cod on the reef compared to locations away from the reef in two out of the three years. In the third year, a possible aggregation of cod due to the reef structure was masked by the effects of a strong year class of sandeel in the vicinity, which led to a similar attraction of cod to the reference area away from the reef (Todd et al. 1992).

With respect to the observations that fish are found in higher densities around artificial reefs, two main explanations have been put forward: the "production hypothesis" and the "attraction hypothesis" (Bohnsack 1989; Svane and Petersen 2001). According to the former, elevated fish densities are caused by the enhanced biomass production on the reef, whereas the latter assumes that aggregations are caused by behavioural responses of fish in the vicinity, which are attracted by the reef. We see a combination of these two processes as the most likely reason for the observed changes.

Structure and Diversity of Fish Assemblages

All of the studies mentioned above are based on estimates of either total abundance of fish, or on a selection of a few fish species of specific regional and/or commercial interest. To our knowledge, no detailed – especially long-term – analyses of entire fish assemblages around either decommissioned oil platforms or already installed wind parks in the North or Baltic Seas have yet been published.

To estimate the potential effects of new wind farms on fish, we are therefore using a combination of the detailed long-term recordings of fish assemblages from the regular survey programs in the German North and Baltic Seas, together with information on the habitat preference of several relevant species. The analysis of North Sea or Baltic fish assemblages in dependence of the structure of the seabed may give indication as to which species would be favoured by the introduction of new hard substrates such as the foundations and pilings of wind turbines.

We hypothesise that in first approximation, effects of newly installed wind parks in the German EEZs and territorial waters will be restricted to the close vicinity of the wind mills, unless rare species are directly affected for which a local reduction causes an overall limitation of breeding success in the population. Otherwise, covering patches of sandy sea floor will probably lead to local removal of species dependent on soft bottom habitats, and will favour hard-bottom fauna and large predators; the latter particularly if the wind parks are closed for fishing activities.

Pihl and Wennhage (2002) analysed the composition of fish assemblages from shallow rocky and soft bottom habitats on the Swedish west coast (Skagerrak). Fifty-three fish species were recorded, of which 30 were common in both habitats. Most of these species also occur in both, the German North and Baltic Seas. In the North Sea, the most abundant species of demersal fish caught throughout the 50-year survey have been the dab (*Limanda limanda*), the whiting (*Merlangius merlangus*) and the plaice (*Pleuronectes platessa*). They all occur on sand as well as on rocky sea floors (Pihl and Wennhage 2002), and hence we anticipate only minor effects on their populations. The three most dominant species in the German Baltic Sea waters are the cod (*Gadus morhua*), the dab and the flounder (*Platichthys flesus*), which again use both soft and hard bottom habitats (Pihl and Wennhage 2002).

Although we do not anticipate that the structural change introduced by the installation of wind turbines (not considering the construction process itself) will have major direct impacts on the populations of most fish species in German waters, long-term effects on smaller spatial scales are likely to be more complex. Their analysis and management will – as has been concluded for artificial reefs – require a whole-ecosystem approach, including long-term analyses of species assemblages, investigation of the mechanisms of species interactions as well as quantification of processes in the biological and physical environments.

References

Bohnsack JA (1989) Are high densities of fishes at artificial reefs the result of habitat limitation or behavioural preference? Bull Mar Sci 44:631-645

Clarke KR, Warwick RM (1994) Change in marine communities: an approach to statistical analysis and interpretation. Plymouth: Plymouth Marine Laboratory, 144 pp

Corten A, van de Kamp G (1996) Variation in abundance of southern fish species. ICES Journal Mar Sci 53:1113-1119

Dippner JW (1993) A frontal-resolving model for the German Bight. Cont Shelf Res 13:49-66

Ehrich S, Stransky C (1999) Fishing effects in northeast Atlantic shelf seas: patterns in fishing effort, diversity and community structure. VI. Gale effects on vertical distribution and structure of a fish assemblage in the North Sea. Fisheries Research 40:185-193

Ehrich S, Stransky C (2001) Spatial and temporal changes in the southern species component of North Sea bottom fish assemblages. In: Kröncke I, Türkay M,

Sündermann J (eds) Burning issues of North Sea ecology. Proceedings of the 14th international Senckenberg Conference North Sea 2000, Senckenbergiana marit 31(2):143-150

Ehrich S, Adlerstein S, Götz S, Mergardt N, Temming A (1998) Variation in meso scale fish distribution in the North Sea. ICES C.M. 1998/J:25, 14 pp

Fricke R, Berghahn R, Neudecker T (1995) Rote Liste der Rundmäuler und Meeresfische des deutschen Wattenmeer- und Nordseebereichs (mit Anhängen: nicht gefährdete Arten). Schriftenreihe für Landschaftspflege und Naturschutz 44:101-113

Heessen HJL, Hislop JRG, Boon TW (1996) An invasion of the North Sea by blue-mouth, *Helicolenus dactylopterus* (Pices, Scorpaenidae). ICES Journal Mar Sci 53:874-877

Hempel G, Nellen W (1974) Fische der Ostsee. In: Magaard L, Reinheimer G (eds) Meereskunde der Ostsee. Springer, Berlin, pp 215-232

Hvid CB, Brünner L, Knudsen FR (2005) Hydroacoustic monitoring of fish communities in offshore wind farms. Annual Report 2004. Horns Rev Offshore Wind Farm. Doc. no 2519-03-003-rev3, 21 pp (http://www.hornsrev.dk/)

ICES (1999) Manual for the International Bottom Trawl Surveys, Revision VI. ICES CM 1999/D:2, Addendum 2

Jensen H, Kristensen PS, Hoffmann E (2004) Sandeels in the wind farm area at Horns Reef. Final report to ELSAM. Charlottenlund: Danish Institut for Fisheries Research, 9 pp. plus Tables and Figures. http://www.hornsrev.dk/Miljoeforhold/miljoerapporter/sandeels_Final%202004_rev_01.pdf

Leewis RJ, Hallie F (2000) An artificial reef experiment off the Dutch coast. In: Jensen AJ, Collins KJ, Lockwood APM (eds) Artificicial Reefs of Europe. Dordrecht: Kluwer, 289-305.

Løkkeborg S, Humborstad O-B, Jørgensen T, Soldal AV (2002) Spatio-temporal variations in gillnet catch rates in the vicinity of North Sea oil platforms. ICES J Mar Sci 59, Suppl:294-299

Pihl L, Wennhage H (2002) Structure and diversity of fish assemblages on rocky and soft bottom shores on the Swedish west coast. J Fish Biol 61 (suppl A): 148-166

Rachor E, Nehmer P (2003) Erfassung und Bewertung ökologisch wertvoller Lebensräume in der Nordsee. Abschlussbericht für das F+E-Vorhaben FKZ 899-85-310 (Bundesamt für Naturschutz), Bremerhaven: Alfred-Wegener-Institut für Polar- und Meeresforschung, 175 pp (http://www.bfn.de/marinehabitate)

Soldal AV, Svellingen I, Jørgensen T, Løkkeberg S (2002) Rigs-to-reefs in the North Sea: hydroacoustic quantification of fish in the vicinity of a semi-cold platform. ICES J Mar Sci 59 Suppl:281-287

Svane I, Petersen JK (2001) On the problems of epibioses, fouling and artificial reefs, a review. Mar Ecol 22:169-188

Todd CD, Bentley MG, Kinnear JAM (1992) Torness artificial reef project. Pp. 15-22. In: Baine MPS (ed) Proceedings of the First British Conference on Artificial Reefs and Restocking, Orkney. ICIT, Orkney, 66p

Valdemarsen JW (1979) Behaviour aspects of fish in relation to oil platforms in the North Sea. ICES C.M. B 27

Research on Benthic Associations

Background

The benthos includes all animals and plants living on the floor of the oceans and inland bodies of water, including both sessile organisms and animals living at the bottom of these waters which creep, walk or temporarily swim. Benthic species are classified as macro-, meio- or micro-benthos, depending on their size.

The conditions of life of the benthos are dependent on a number of abiotic factors, such as sediment conditions, salt content, light conditions, temperature and depth of water. The activities of the zoobenthos are limited largely to the interface area between the free water and the topmost layer of soil. The species composition and the numbers of individuals are subject to major fluctuations over time, depending on various factors from food supply over to currents and water temperature.

In the German marine areas of the North and Baltic Seas, the benthic habitat is strongly subject to anthropogenic burdens. Especially the heavy beam trawls used in bottom fisheries mix up the upper layer of the ocean floor and its benthic fauna extensively.

The effects of offshore wind-energy plants on the benthos are still largely unexplored. The construction of offshore wind parks could, through structural installation over the seabed and shifting of sediments due to construction measures, affect the benthic communities or individual species. Non-mobile or hardly mobile species and suspension-feeding species (filterers) are particularly susceptible. The construction of the foundations introduces artificial hard substrata to the ocean floor which will attract specific benthos species, resulting changes in the benthic associations at sandy locations. Due to the central function of the benthos in the marine ecosystem, it cannot be ruled out that the introduction of artificial hard substrata and the resulting settlement by different benthos associations could result in a change or a shift of the entire species spectrum, and hence of the natural biocoenosis.

In summary, the following correlations of effects are relevant with regard to the consequences for the benthic associations:

- Spatial demands by sediment shifts and the installation of the main body of the facility, causing the elimination of benthic associations or single species;
- Changes in species composition due to the introduction of artificial hard substrata, and through shifts in sedimentation and hydrodynamic conditions;
- A rise in sediment temperature in the vicinity of electric cables.

As in the case of fish, a fishing ban will affect the benthos associations within the wind park area as well. Initial indications have supported assumptions that particularly large and mostly fragile species occur in the "protected" areas. Knowledge about possible effects of electromagnetic fields (e.g. impairment of orientation, avoidance behaviour) on benthic associations is lacking. There is no knowledge whatsoever on the results of temperature rises in the direct vicinity of the cables.

Existing knowledge does not give rise to any expectation that the construction and operation of offshore wind parks might cause any considerable impairments which might lead to a denial of their approval. Nevertheless, existing uncertainties must be cleared up, and existing questions answered, particularly regarding the long-term effects on benthic associations. The need for research therefore exists primarily in the area of the examination of the progression of settlement and of the species stock at the piles of the plants, and the results of that process. In addition to this question, the research projects introduced in this chapter also address the changes in hydrodynamic conditions in the areas around the piles, the resulting changes in sediment and the fauna, and the effects of electromagnetic fields on invertebrates.

12   Benthos in the Vicinity of Piles: FINO 1 (North Sea)

Alexander Schröder, Covadonga Orejas, Tanja Joschko

Changes in the macro-zoobenthic communities in the areas of offshore wind farms are the result of the cumulative effects of numerous piles on the marine fauna. The essential processes take place in the vicinity of these piles. An altered hydrographical regime leads to erosion and to changes in sediment composition in the direct surrounding area of the piles. In addition, the underwater structure provides an artificial hard substrate for many organisms which rarely occur in the typical soft bottom communities of the German Bight. These also influence the food supply for the fauna below and attract predators, which in turn may cause changes of the bottom communities by increasing the predatory pressure.

Until very recently, the influence of wind farms on the marine environment could only be estimated theoretically on the basis of the results of investigations of other artificial hard substrates, such as wrecks or artificial reefs. These however represent different structures of different dimensions. Initial results from an offshore wind farm at Horns Rev Denmark, have been presented (ELSAM 2004a, 2004b), but are only partially transferable to the planned installations in the German Bight, since both the hydrographic and the sedimentological conditions are considerably different.

The possibilities of investigating these processes directly at the planned wind energy plants are somewhat restricted, for technical and logistical reasons. Process-oriented studies are for these reasons not part of the standard monitoring programme during construction and operation of such installations. The research platform FINO 1[1] therefore provided an ideal possibility to study these processes in depth at the piles, prior to the actual construction of the wind farms.

Within the framework of the BeoFINO[2] research project, several studies addressed questions of the possible ecological effects of offshore wind farms (see also chapters 9 and 13). It is hoped that the results can help assure the environmentally compatible construction of the planned offshore wind farms.

---

[1] „Forschungsplattformen in Nord- und Ostsee" = research platforms in the North and Baltic Seas

[2] Ecological research into offshore use of wind energy, on research platforms in the North and Baltic Seas (BeoFINO), funded by the German Federal Ministry for the Environment (BMU)

12.1   Marine Ecological Research at the FINO 1 Platform

The scientific research platform FINO 1 was installed in July 2003 in the German Bight, at 54°00'52" N, 006°35'16" E, approx. 45 km north of the island Borkum[3]. Situated in 28 m water depth on a bottom of homogeneous fine sand, it was intended to measure various factors which play an important role for the planning and operation of offshore wind farms. In addition to various physical measurements, such as wind velocities and stress caused by waves, it also hosts a number of studies investigating possible environmental effects.

The platform rests on a jacket structure with four braced piles, spreading from a square of 7.5 m at the surface to one of 26 m at the anchorage to the sea bed (see Fig. 3 in chap. 15). Hydrographical measurements are carried out by the German Federal Maritime and Hydrographic Agency (BSH), which has kindly supplied these data for our analysis. Current velocities at the platform are mainly dominated by tidal currents, and reach 1 m/s at the surface and approx. 0.4 m/s near the bottom twice daily.

A remote-controlled digital underwater camera was installed at one pile of the platform to allow continuous documentation of settlement and succession processes at a fine time scale over the entire depth. Quantitative epifauna samples from the underwater structure were taken by divers to allow detailed taxonomic identifications and quantification of biomass.

A crane allows sampling at various distances from the research platform, up to 15 m away from the underwater structure. Infaunal samples were taken by grabs, and sediment samples were analysed for grain size composition and organic content. Reference samples were taken from shipboard within a wider area, approx. 200 - 400 m around the platform. The addition of artificial hard substrates was not the only effect of the installation of the platform. As in the case of all offshore structures (including wind farms), a radius of 500 m is closed to normal shipping, for safety reasons. As this includes fisheries, the area is also protected from bottom trawling, which is very intensive in this part of the German Bight. To permit a distinction between the direct platform effects and those of fishing closure, the reference samples were taken within the 500 m safety-area in which fishing was excluded.

The main goals of our study are the documentation of possible alteration of sediment quality due to hydrodynamic changes close to the piles, and

---

[3] see also: www.fino-offshore.de

the description, quantification and interpretation of the settlement, recruitment and development processes of the benthic communities on both soft bottoms and on the pile itself as artificial hard substrate. The results presented here mainly cover the period from March 2003 – before construction – to December 2004.

12.2   Fauna on Artificial Hard Substrate

The underwater structure of the FINO 1 platform provides an artificial hard substrate which forms a new habitat for marine epifaunal organisms, which rarely occur in soft bottom communities, such as some sea anemones (Actinaria) and mussels (Bivalvia) (e.g. blue mussels). Situated in an area of only sandy bottoms, the nearest natural habitats with hard substrates are more than 20 km away. However, numerous wrecks in the German Bight may constitute similar structures inhabited by a dense epifauna, and act as stepping stones for the colonisation of new substrates. Many marine organisms reproduce and disperse by means of planktonic larvae. Colonisation thus depends on the one hand on the available larval supply, and hence on the origin of the water masses passing the structure, and on the other on the suitability of the substrate. In addition to its isolation from other hard substrate habitats, the platform also constitutes a special habitat itself, as it spans the entire range of water depth, and is greatly exposed to currents and waves. Moreover, the material of the platform (steel) differs from natural substrates. These factors may explain why the epifauna of the platform has differed from that on natural marine hard substrates in the German Bight, e.g. the rocky grounds around Helgoland, not only in its species composition but also in the dynamics of its faunal succession.

12.2.1 First Arrivals

Very soon after construction of the platform, the marine fauna began to occupy the new grounds. Numerous organisms colonised the underwater structure rapidly and densely. This fast and intense process is typical for the first phase of an ecological succession process (e.g. Connell and Slatyer 1977). The order of appearance depended on the time of the installation and the supply of larvae ready to settle. Amongst these are the larvae of the "pioneer species" at FINO 1: Only two weeks after the construction, the hydroid *Ectopleura larynx* almost completely covered the surface of the underwater structure (Fig. 1). Hydroids are well known pioneer species in many marine environments, due to their adaptability to various sub-strates and their fast life cycles (Gili and Hughes 1995).

In many cases, the establishment of the first species will determine the next arrivals: At FINO 1, a nudibranch (*Coryphella browni*) was the second inhabitant, feeding on the hydroid *Ectopleura larynx*. This initiated an ongoing succession process, with more and more species arriving. Upon arriving at the platform, their settlement depended on the suitability of the substrate and on their ability to compete with those already there.

Fig. 1. Dense cover of hydrozoans (*Ectopleura larynx*) on the underwater struc-ture in August 2003

12.2.2 Fighting for Space

Following this rapid initial settlement, various new settlers arrived one after another and competed for space. Only the faster and stronger ones could be successful. Space seems to be the most limiting resource in the colonisation process at FINO 1. Competing for space, growing over one another, fighting for food, living as permanent or seasonal inhabitants or just as accidental visitors – the epifaunal picture on the underwater surface of FINO 1 changed continuously.

After some time, the pile was in large parts densely covered by the amphipod *Jassa herdmani* (former syn. *J. falcata*), which builds dense tube mats accumulating to a large biomass (Fig. 2). From the summer of 2004 onwards, the higher sections (up to approx. 5 m depth) of the piles were dominated by the blue mussel (*Mytilus edulis*), which provides a rich food source for the subsequently appearing predatory starfish (*Asterias rubens*).

Fig. 2. Coverage of the surface by tube mats of the amphipod *Jassa herdmani*

In addition to space competition, predation plays an important role in the development of the community. Furthermore, the seasonal occurrence of certain species strongly determines the appearance of the epifaunal community. Some species, like the hydroid *Ectopleura larynx,* not only appear early on empty grounds, but also reappear again every summer, settling even on the existing fauna. However, at other times, other species take over and became dominant at certain depths, giving rise to a highly dynamic pattern of succession.

Compared to natural hard substrates and coastal habitats, the species spectrum found to date on the piles shows rather low diversity (Fig. 3). However, the present situation has not yet reached a steady "climax" state, and further arrivals are still expected.

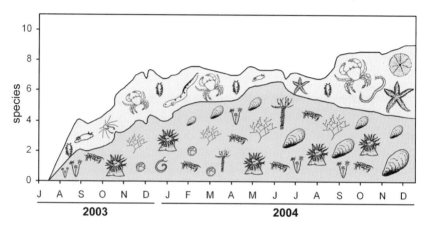

Fig. 3. Temporal development of epifaunal species number on FINO 1. The lower area shows the mean number of mostly sessile species. The addition of mobile organisms sums up to the total number of species per photograph (approx. 0.04 m²)

12.2.3 The Shallow and the Deep

Initially, the entire surface was covered by a very simple and quite homogeneous community, differing only in slightly varying densities of organisms. Over time, three main depth zones (0 - 5 m, 5 - 25 m, 25 - 29 m) became discernable, each with a characteristic faunal composition (Fig. 4). During the second year, the higher sections up to 5 m depth became completely covered by mussels (*Mytilus edulis*); their densities decreased with increasing depth, and below 5 - 6 m, only single mussels lived within a dense cover of hydrozoans and amphipods. The amphipod *Jassa herdmani* built a dense mat of tubes covering all available surfaces. This tube mat grew very quickly and in times reached a thickness of more than 5 cm. In these mats, *Jassa herdmani* reached densities of more than 2.4 million individuals per m². In between them live several sea anemones, such as *Metridium senile* and, especially in the lower reaches, also *Sagartiogeton undatus*. Predators like the common starfish (*Asterias rubens*) and the swimming crab (*Liocarcinus holsatus*) roam the surface foraging for food. Near the bottom there are also some smaller edible crabs (*Cancer pagurus*)

0 - 3 m

3 - 5 m

5 - 10 m

10 - 25 m

25 - 29 m

climbing the piles, while larger adult individuals are found on the ground around the piles.

The largest accumulation of biomass consisted, in the upper reaches, of a dense cover of mussels, and in the lower reaches, mainly of hydrozoans, amphipod tube-mats and large sea anemones (*Metridium senile*). While the mussels form a relatively stable and compact cover, the hydrozoans appear seasonally, and the amphipod tube mats, partly growing on hydrozoans, are also constantly being torn apart and rebuilt. Near the bottom, the cover becomes thinner and less diverse. Some suspension feeders (e.g. hydroids and bryozoans) did not appear in the deepest areas, because the suspension of fine sediments near the bottom hindered the efficient filtration of the organisms (e.g. Reiswig 1973, Round et al. 1961).

Fig. 4. Depth gradient of epifauna on FINO 1 in Nov. 2004

12.2.4 Accumulation of Biomass

The settlements of numerous organisms from various species lead to an enormous accumulation of biomass on the underwater structure, with approx. 3.6 t in the summer of 2004. The upper areas were covered by small (1 - 2 mm) mussels, but the biogeneous layer was not as thick as in deeper parts, where an average biomass of almost 3.5 kg/m² was found (Fig. 5). Although the samples from 2005 have not been completely analysed yet, a much larger biomass was observed during the sampling in the upper reaches. Mussels of approx. 5 cm length have formed a massive layer of 10 - 15 cm thickness, and it can already be stated that the biomass is much higher than in 2004.

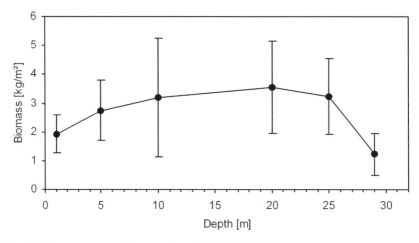

Fig. 5. Biomass accumulated on the pile in July 2004

The blue mussel (*Mytilus edulis*) constitutes most of the biomass in the shallower areas of the piles, while the amphipod *Jassa herdmani* dominates in the deeper zones. A part of this biomass is constantly eroded by wave action or movements of mobile predators, and, together with the faeces of all organisms, sinks to the ground. Due to this additional input of organic matter, the food supply to epifaunal predators and the organic matter content of the sediments increases around the piles. The amount of exported matter is very difficult to estimate at present, but will be addressed in future research.

This enormous amount of biomass provides food sources not only for benthic organisms, but also for such higher predators as fish. Near the platform, large aggregations of horse mackerel (*Trachurus trachurus*) have been observed (Fig. 6), and around the feet of the piles, some gadoids have been seen by divers. This phenomenon is well-known from investigations on oil-platforms in the North Sea (e.g. Løkkeborg et al. 2002), where an attraction of some fish species by artificial underwater structures was documented.

Fig. 6. Aggregation of horse mackerel (*Trachurus trachurus*) around the platform

12.3   Soft Bottom Fauna

The natural sea bottom in most areas proposed for offshore wind farms in the German Bight consists of soft sediments ranging from muddy fine sand to coarse sand (Fig. 7). These sediments are inhabited by an adapted fauna, most of which lives beneath the surface and is not visible at first glance. Nevertheless, every square meter is inhabited by a large numbers of organisms.

Fig. 7. Typical appearance of fine sandy bottoms

Many benthic organisms are sessile, and can hardly escape natural and anthropogenic changes. Several species are known to be susceptible to disturbances or environmental changes. Because of its mostly sessile character and its ability to "integrate" environmental influences over longer time scales, the macro-zoobenthos is commonly regarded as a good indicator for environmental impacts (Underwood 1996), as well as for long-term changes in the ecosystem (Kröncke 1995). It is – especially in shallow shelf seas – an integral part of the system, with major importance in the remineralisation and transformation of deposited organic matter (Josefson et al. 2002), and constitutes the main food resource for demersal fish (Reid 1987).

In the vicinity of hard substrates, modified hydrographic regimes alter sediment properties and hard substrate fauna influences the surrounding soft bottom fauna both directly and indirectly.

12.3.1 Alterations of Sediments

The construction of FINO 1 has changed the hydrographic regime in the direct surrounding of the platform, resulting in significant changes in local sediment composition. Artificial underwater structures can change the physical conditions in their surroundings, such as local current speeds and organic carbon contents, thereby creating altered sediment conditions (e.g. Davis et al. 1982, Ambrose and Anderson 1990). Close to the FINO 1 platform (up to 5 m), the sediment is much more heterogeneous than it was prior to construction. It contains many more shells, probably washed out of the sediment by erosion. In the direct vicinity of the piles which are fixing the platform in the ground, increased bottom current velocities lead to deep scours of approx. 1 to 1.5 m depth. While lighter mobile sands are resuspended and transported away, dead shells remain on the ground and form a layer of sometimes more than 30 cm thickness. Further away from the platform (200 - 400 m), no changes in sediment composition were observed.

12.3.2 Changes in Faunal Communities

From the outset, the fauna in the direct vicinity of the platform was altered, most prominently at 1 m distance from the pile. This close to the platform, there was surely a direct influence of the construction works, causing a diminishment of the fauna; however the alterations of the sediments also influenced the ability of many species to colonise this area. The temporal and spatial variability of benthic species depends among other things on the survival rate and successful recruitment of juveniles (Kuenitzer 1992). As settlement and recruitment of larvae are related to sediment quality, several species did not settle in this area, while others were not able to establish themselves permanently after initial settlement. Many typical soft bottom inhabitants were absent or much less abundant around the platform. This particularly affected species which live burrowed in the ground as the bivalve *Tellina fabula* or the heart urchin *Echinocardium cordatum*, but also the Ophiuroids (Fig. 8a), which are normally quite abundant in such communities. Many polychaetes, which constitute a major component of typical soft bottom fauna, were found, if at all, only in very low densities in the vicinity of the platform, such as *Poecilochaetus serpens*, *Chaetozone setosa*, or such tube building species as *Spiophanes bombyx* (Fig. 8b).

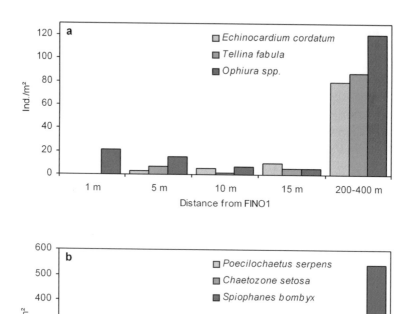

Fig. 8. Density of typical soft bottom fauna around FINO 1 in the second half of 2004. **a**: Echinoderms and Bivalves; **b**: Polychaetes

However, a few species did reach higher densities around the piles than in the reference areas. Mobile predators, such as hermit crabs (*Pagurus bernhardus*), swimming crabs (*Liocarcinus holsatus*), some amphipods and carnivorous polychaetes, such as *Eunereis longissima*, appeared in larger numbers in nearby areas (Fig. 9).

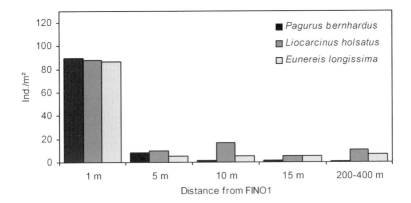

Fig. 9. Density of predatory species around FINO 1 in the second half of 2004

A combination of various factors including changes in sediment structure in conjunction with increased predatory pressure, resulted in the displacement of typical soft bottom species. Currently, the effect of the research platform on the benthic community still seems to be spreading further, and is already noticeable at up to at least 15 m distance from the platform. The spatial and temporal extension of these changes and their consequences for e.g. trophic interactions are the major issues in the ongoing studies.

However, the addition of artificial hard substrates was not the only effect due to the construction of the platform. The closure of a radius of 500 m to fisheries protected the area from bottom trawling, which is very intensive in this area. This alone led to some initial changes in the dominance structure of the benthic community compared to the normally fished areas around the platform. These changes were investigated within the frame of the EU Project RESPONSE (final report in press). Mobile predators and scavengers reached higher densities in the fished areas, especially during the main fishing seasons, as these are attracted by damaged and dead organisms hit by the trawl gear (e.g. Groenewold and Fonds 2000). The initial recovery of the protected area during the first year after the installation of the platform resulted in increased densities of many sedentary filter and deposit feeders. Yet these were still mainly opportunistic species; a possible development towards a community with larger, longer-lived species would take much more time.

12.4   Conclusions

The intensive investigations at the research platform FINO 1 over a period of 18 months have produced numerous results, which permit some initial conclusions. However, this constitutes only the first phase of the reaction of the fauna to the introduced structure, and several long-term aspects cannot yet be conclusively assessed.

A highly dynamic epifauna has settled on the underwater structure, reaching an as yet relatively low diversity, albeit with a high biomass. With its strong seasonal fluctuations, it produces a continuous export of biogenic material, which is hard to quantify at present. The lumps of biomass sink relatively quickly to the ground; they may be shifted to an as yet unknown distance by the currents but will mainly concentrate around the pile. This material not only provides additional resources, but more importantly a different quality of food. The usual food web is based on small particulate matter which settles to the ground and is consumed by filter and detritus feeders in the common soft bottom fauna. Larger chunks falling off the artificial hard substrate are only available for larger predators and scavengers, and thus present a shortcut within the food web.

At the research platform FINO 1, the surrounding sediments as well as the soft bottom community have already experienced significant alterations. However, equilibrium has not been reached, and the final extension of this sphere of influence cannot yet be estimated definitively. For such a single structure, this effect may be seen as a restricted local phenomenon, but the proposed wind farms, with hundreds of similar structures, may cumulatively lead to considerable impacts. As present, results indicate that significant impacts can be seen from even a single structure; the uncertainty as to the spatial extent and ultimate quantitative and qualitative change makes predictions regarding larger wind farms very speculative. The effects will also depend on the type of underwater structures, the materials used, and the question as to whether anti-fouling or scour protection is applied or not. These issues have not yet been decided.

In an environment like the German Bight, where the fauna is generally dominated by short living opportunistic species (Wieking and Kröncke 2003), these have the greatest chance of reaching a dominant position when a suitable habitat becomes available. On the other hand, the present benthic fauna of the German Bight is not only well adapted to a variable environment with a considerable elasticity against short-term and localised disturbances (Schröder 2005); it has also been shaped by numerous anthropogenic effects for a long time, and is probably altered continuously by chronic but patchy trawling (Jennings et al. 1999). It is often not possible to distinguish the influence of a single factor from the multiple influences of e.g. pollution, eutrophication and fisheries (Rachor and Schröder 2003). Some aspects, such as the promotion of predators and scavengers by additional food supply from bio-fouling at wind farms and from fishing, may work in the same direction. Others may work in opposite ways, making predictions about the direction of major changes due to wind parks very difficult.

Real changes in the community composition cannot be expected in the short term, or on small spatial scales. A possible succession towards a more mature community of long-living larger species will require a much longer time span and a larger scale. A stable situation has not yet evolved, and the further development as well as an assessment of the cumulative effects of numerous piles composing the wind farms, will be followed in ongoing studies within the BeoFINO II project.

References

Ambrose RF, Anderson TW (1990) Influence of an artificial reef on the surrounding infaunal community. Mar Biol 107:41-52

Connel JH, Slatyer RO (1977) Mechanisms of succession in natural communities and their role in community stability and organization. Am Nat 111:1119-1144

Davis N, VanBlaricom GR, Dayton PK (1982) Man made structures on marine sediments: Effects on adjacent benthic communities. Mar Biol 70:295-303

ELSAM (2004a) Horns Rev. Infauna Monitoring. Horns Rev Offshore Wind Farm. Annual Status Report 2004

ELSAM (2004b) Horns Rev. Hard Bottom Substrate Monitoring. Annual Status Report 2004

Gili JM, Hughes RG (1995) The ecology of benthic hydroids. Oceanogr Mar Biol Ann Rev 33:351-426

Groenewold S, Fonds M (2000) Effects on benthic scavengers of discards and damaged benthos produced by the beam-trawl fishery in the southern North Sea. ICES J Mar Sci 57:1395-1406

Jennings S, Alvsvaag J, Cotter AJR, Ehrich S, Greenstreet SPR, Jarre-Teichmann A, Mergardt N, Rijnsdorp AD, Smedstad O (1999) Fishing effects in northeast Atlantic shelf seas: Patterns in fishing effort, diversity and community structure. III. International trawling effort in the North Sea: An analysis of spatial and temporal trends. Fish Res 40:125-134

Josefson AB, Forbes TL, Rosenberg R (2002) Fate of phytodetritus in marine sediments: functional importance of macrofaunal community. Mar Ecol Prog Ser 230:71-85

Kröncke I (1995) Long-term changes in North Sea benthos. Senckenb Marit 26:73-80

Kuenitzer A (1992) Does settlement influence population dynamics of macrobenthos? A case study in the central North Sea. In: Marine eutrophication and population dynamics, with a special section on the Adriatic Sea. 25th European Marine Biology Symposion., Olson and Olson, Fredensborg (Denmark), Int Symp Ser pp 285-292

Løkkeborg S, Humborstad OB, Jorgensen T (2002) Spatio-temporal variations in gillnet catch rates in the vicinity of North Sea oil platforms ICES J Mar Sci 59:294-299

Rachor E, Schröder A (2003) Auswirkungen auf das Makrozoobenthos - Nutznießer und Geschädigte der Eutrophierung. In: Lozán JL, Rachor E, Reise K, Sündermann J, von Westernhagen H (eds.) Warnsignale aus Nordsee and Wattenmeer. Eine aktuelle Umweltbilanz. Verlag Wissenschaftliche Auswertungen, Hamburg. pp 201-203

Reid PC (1987) The importance of the planktonic ecosystem of the North Sea in the context of oil and gas development. Phil Trans R Soc Lond Ser B 316B:587-602

Reiswig HM (1973) Population dynamics of three Jamaican Demospongiae. Bull Mar Sci 23:191-226

Round FE, Sloane JF, Ebling FJ, Kitching JA (1961) The ecology of Lough Ine. X. the the hydroid *Sertularia operculata* (L.) and its associated flora and fauna: effects of transference to sheltered water. *J Ecol* 49:617-629

Schröder A (2005) Community dynamics and development of soft bottom macrozoobenthos in the German Bight (North Sea) 1969 - 2000. Ber Pol Meeresf 494: 181 pp

Wieking G, Kröncke I (2003) Macrofauna communities of the Dogger Bank (central North Sea) in the late 1990s: spatial distribution, species composition and trophic structure. Helgol Mar Res 57:34-46

Underwood AJ (1996) Detection, interpretation, prediction and management of environmental disturbances: Some roles for experimental marine ecology. J Exp Mar Biol Ecol 200:1-27

13 The Impact of Wind Engine Constructions on Benthic Growth Patterns in the Western Baltic

Michael L Zettler and Falk Pollehne

13.1 Introduction

Global-scale environmental degradation and its association with non-renewable fossil fuels have led to an increasing interest in generation of electricity by renewable energy resources (Gill 2005). Since the planning of large offshore wind energy facilities in the German Bight and the Baltic Sea was initiated, concerns about the ecological compatibility of these structures have been expressed. Apart from direct impacts of disturbance during construction, operational sounds and rotating parts, which might primarily affect birds, bats, marine mammals and fish, the potential long term effects on the benthic environment have been discussed. These concerns are mainly focused on the questions, whether and how the natural benthic habitat in the vicinity of the constructions is modified by changes in bottom currents and turbulence, and whether the effects of the installations as artificial settling substrates are properly assessed. The ecologically relevant effects of offshore wind parks include e.g., increased habitat heterogeneity, and changes in hydrodynamic conditions and in sediment transport patterns. The potential ecological response of the macro-zoobenthos could involve long-term changes in diversity, abundance, biomass, community structure and such functional properties as nutrient regeneration or bio-turbation.

These problems have been in the focus of a project in the western Baltic which that was part of a national combination of projects called BeoFINO.[1] This effort has addressed the overall ecological risks of offshore wind-power facilities in the North and Baltic Seas.

Such questions are most often viewed in the primary context of the effects on the biodiversity of the benthic community. In the Baltic Sea however, the specific hydrographical conditions emphasizes a problem which also involves the absolute biomass accumulation rates of epifauna

---

[1] Ecological research on offshore use of wind energy on research platforms in the North and the Baltic Seas (BeoFINO), established by the German Federal Ministry for the Environment, Nature Conservation and Nuclear Safety.

on substrates that protrude into the surface mixed layer. Particularly in the inflow areas of denser, more saline North Sea water adjacent to the Belt Sea and the Danish Sound, severe vertical stratification between the surface mixed layer and the bottom water overlying the sediments is the rule rather than the exception. The stratification is much more stable than in the North Sea, as tidal mixing is not an effective source of vertical exchange in the Baltic. Surface productivity is high in these areas, at least partly due to anthropogenic eutrophication, and as the density gradient does not constrain organic particles from sinking into deeper water, but prevents dissolved oxygen from mixing downwards, these benthic areas are extremely susceptible to oxygen deficiency. The increase of benthic biomass due to enhanced nutrition over the past 50 years (Karlson et al. 2002) has already aggravated the problem of unbalanced oxygen supply and consumption. In Baltic estuaries with a similarly strong stratification regime, bottom anoxia events have been documented (e.g. Powilleit and Kube, 1997), with destructive wide-ranging effects on benthic ecosystems and such associated economies as fishing, tourism and recreation. Additional point sources of organic matter to the sedimentary systems in such areas may initiate local cores of anoxia, which then start to spread over larger areas in an exponential fashion, when the suffocated benthic biomass is itself subject to microbial decomposition and oxygen demand.

This scenario is particularly alarming, as most projected wind parks in the western Baltic are planned to be positioned exactly in the areas of most intense vertical stratification, either in the Pomeranian Bight at the estuarine stratification of the Oder plume, or at Kriegers Flak at the outlet and subduction area of dense saline water from the Danish Sound. As these environments are extremely sensitive to the input of additional organic matter, the export of benthic biomass from the higher parts of structures to the surrounding sediments became a relevant aspect in the study.

Recent studies on ecological impacts of offshore wind farms on the benthic ecosystems are rare and mainly published as reports (e.g. Birklund and Petersen 2004, Leonard and Pedersen 2004 et al. cited in Gill 2005). Within the present study, both qualitative and quantitative aspects of benthic growth dynamics in the western Baltic at an artificial pile model were investigated. A delay in the construction of a full size research platform in the key area of Kriegers Flak led to the installation of a reduced size model pile in the region of Darss Sill, which is an area restricted to research. Over a period of two years, larval settling dynamics, biomass development and a succession of benthic organisms was observed at and around this pile, as well as on additional artificial settling substrates throughout the water column. The presence of an adjacent autonomous monitoring station that registers and logs such basic environmental data as salinity, temperature and currents supported the interpretation of the results.

13.2 Material and Methods

13.2.1 Investigation Area

As the ecological and faunistic background of the Baltic is totally different from that of the North Sea, a simple translation of results between the two research areas would not be plausible, so that basic investigations were carried out in both areas. While the North Sea group (Alfred Wegener Institute, Bremerhaven) was able to observe the process of primary colonisation at the full size research platform FINO 1 off Borkum Riff in the eastern German Bight, construction of the Baltic platform was delayed and a model steel substrate was submersed at the Darss Sill station at 54° 41.764' N; 12° 42.085' E, at a water depth of 20 m (Fig. 1).

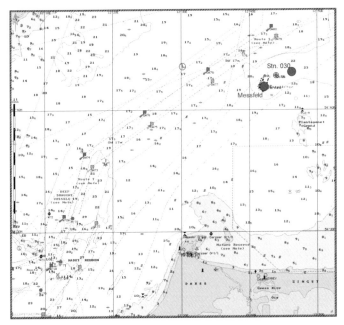

Fig. 1. Position of the experimental field at IOW-Messfeld (red). Stn. 030 (blue) marks a position, where a long term series of zoobenthos data from the HELCOM program is available

13.2.2 Design of the Sampling

Due to continuous high bottom currents driven by North Sea water inflow, the sediment is well sorted fine sand (median grain size ~200 µm) and the water column displays a haline stratification throughout most of the year. As a model for the base piles of offshore wind turbines, an uncoated steel cylinder of 2.2 m in diameter and 2 m in height (Fig. 2) was installed at a water depth of 20 m. The material was selected following the guidelines of the planning authorities for the construction of offshore turbine basements. The changes in hydrodynamic conditions close to the cylinder were recorded by an uw-video system, visual inspections and sampling by divers.

Fig. 2. Steel segment, which was deployed in April 2003 at 20 m depth to simulate the foot area of an offshore wind engine pile

Three additional moorings with square steel tiles of 20 x 20 cm at five different water levels throughout the water column (3, 6, 9, 12 and 15 m above bottom, Fig. 3) were deployed at the start of the experiment in April 2003, and were recovered consecutively after 143, 246 and 470 days. Each recovery of the substrates was accompanied by scratch sampling of an area of 20 x 20 cm at the main steel cylinder 1 m above ground by scuba divers. In July 2004, three new sets of steel tiles were moored, and the first one retrieved in January 2005 after 177 days, in order to study the dependence of the colonisation dynamics on the seasonal phase of deployment.

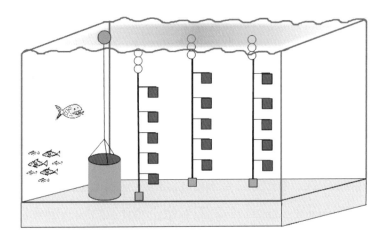

Fig. 3. Scheme of the experimental setup at the Darss-Sill station. On the left, note the permanently exposed steel cylinder, and on the right, the periodically exchanged steel settling substrates. Water depth is 20 m; the vertical distance between the artificial substrates is 3 m

Sampling of the ambient sediments was performed by means of by diver-operated acrylic cores of 10 cm diameter and 50 cm length. The penetration depths of these cores, and hence the mean sampling depth, was 15 cm. In May 2004 and March 2005, samples were taken at distances of 0.5; 1.0; 1.5; 2.0 and 2.5 m in all four directions from the cylinder, to estimate the range of impact (sampling scheme see Fig. 4).

The samples were sieved through 500 μm mesh and preserved in 4 % formaldehyde until analysis in the lab. For a comparison and classification of the results, the data from the nearby monitoring station were consulted. This station has been sampled annually for fifteen years, using a 0.1 m² van Veen grab. Samples are sieved through 1 mm mesh; preservation and laboratory analyses are identical to those in the present study.

Temperature, salinity, current speed and direction were recorded at the adjacent automatic observatory (Marnet-Station) of the Federal Maritime and Hydrographic Agency (BSH) which is operated by the Baltic Sea

Research Institute. Data were continuously recorded and kindly supplied by the instrumentation group of the institute.

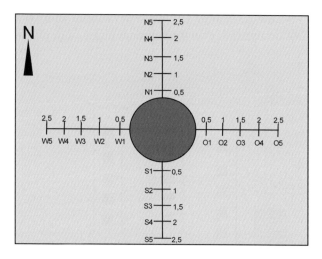

Fig. 4. The diver operated sediment sampling in May 2004 and March 2005 after 417 and 697 days of exposure followed the displayed scheme. The samples were taken at intervals of 0.5 m away from the central steel cylinder. Samples were labelled according to distance and direction

13.3 Results

13.3.1 Hydrographical Boundary Conditions

The time series of temperature and salinity (Fig. 5) show clear different hydrographical backgrounds in 2003 and 2004. Bottom salinity remained in the range of about 20 psu (practical salinity unit) over nearly the whole of 2003, whereas in 2004 larger periods of complete mixing with less saline water in the bottom water occurred. In 2003, one of the largest inflow events of North Sea water into the Baltic was recorded (Feistel et al. 2003), which led to a renewal of bottom water in all basins of the western and central Baltic Sea, and is visible in this dataset as well. As a result of a more intense stratification in 2003 and a generally warm summer, the surface layer could absorb more energy, so that the temperature conditions between these years also differed.

These generally differences in the physical background, with an almost continuous stratification in 2003, were also reflected in the settling and growth dynamics of benthic organisms.

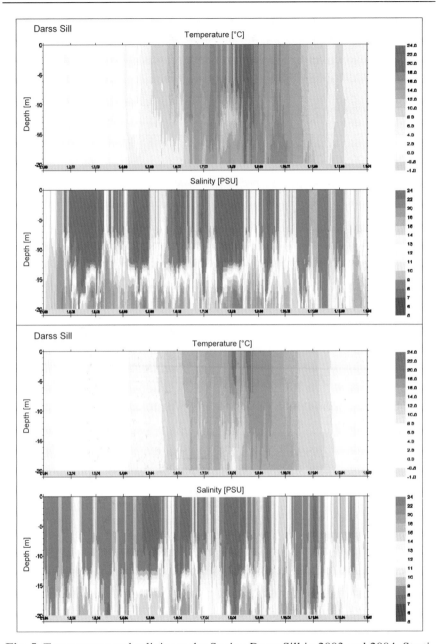

Fig. 5. Temperature and salinity at the Station Darss Sill in 2003 and 2004. Stratification of the water column and heating of the surface layer was much more pronounced in 2003 than in 2004 (data by courtesy of IOW-MARNET Group)

The measurements of the current direction showed a clear alignment in either the south-west or the north-east direction (Baltic outflow or inflow situations, respectively). Mean current speed was in the range of 20 cm sec^{-1} with maximal values reaching 60 cm sec^{-1}.

The recorded long term current conditions can directly be related to the sediment relocation processes at the foot of the cylinder (Fig. 6).

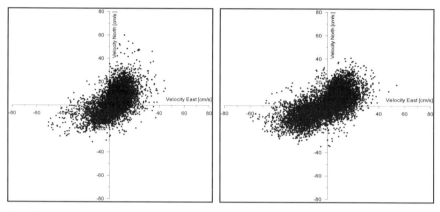

Fig. 6. Direction and speed of near-bottom currents in 2003 (left graph) and 2004 (right graph)

13.3.2 Colonisation of the Basement Model Substrates by Epifauna

After 143 days of exposure, 18 species of macro-fauna could be detected on the steel substrates, of which crustaceans and molluscs dominated, with six and five species, respectively. Less frequent were hydrozoans, polychaetes and echinoderms. Species number was highest (10) close to the bottom. The two main species were barnacles (*Balanus improvisus, Balanus crenatus*) with between 8,000 and 20,000 individuals per m^2, and blue mussels (*Mytilus edulis*), with 70,000 to 470,000 individuals per m^2.

The biomass was also dominated by these two species. While *Mytilus edulis* attained biomasses of between 1.7 and 22.3 g/m² ash free dry weight (afdw), due to the fact that the organisms had settled relatively recently, *Balanus* ssp. grew up to between 8.8 and 146 g afdw/m², peaking at the top plate at 5 m water depths with 169 g afdw/m² and 400,000 individuals. Figure 7 shows the water depth dependent development of biomass and abundance (no. of individuals) during different periods.

After an exposure of 246 days, 28 taxa were recorded, dominated by crustaceans, polychaetes and molluscs. Species number was highest at the base (22 species), and decreased towards the surface (10 species).

Figure 7 depicts the vertical distribution of biomass and abundance after this period. *Mytilus edulis* was the most abundant organism, with between 265,000 and 670,000 individuals per m², increasing by a factor of two as compared to the first sampling. It overgrew the barnacles, which eventually decreased in numbers. The biomass of the blue mussel increased to up to 1 kg afdw/m² in the surface layer. Barnacle biomass increased as well, but remained a factor of 10 lower than the mussels.

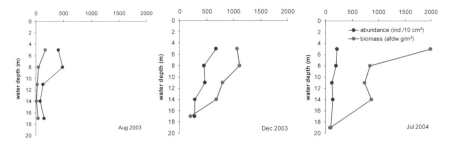

Fig. 7. Development of abundance (individuals per 10 cm²) and biomass (afdw per m²) after 143 (left), 243 (middle) and 470 (right) days of exposure of steel plates in different water depths

Fig. 8. Settling patterns on the bottom cylinder after 264 days. The starfish (*Asterias rubens*) becomes more frequent on the cylinder walls (1-2 individuals per m²). The left photograph shows that reddish rust flakes have formed, and are starting to peel from the wall, with the attached organisms. The right picture shows the basement of the cylinder, where the currents have created a trough by sediment erosion which has started to fill with the debris from the cylinder walls, thereby attracting large predators, such as crabs (*Carcinus maenas*) and starfish. The larger shells of some mussels have been washed out of the sediment

It is evident, that the settling density, abundance and biomass are much higher in the surface mixed layer then at the base of the installation. This is probably also effected by increasing predation pressure by large predators like *Asterias rubens* and *Carcinus maenas*, as can be seen in Fig. 8.

The values in the bottom water levels are lower by a factor of five. The overall colonisation numbers are extremely high, considering the short exposure time. Maximal abundance after 246 days is in the range of 700,000 individuals m², with a biomass of 1.1 kg afdw, equivalent to about 18 kg of wet weight per m².

In July 2004, the succession on the bottom steel cylinder was further advanced. There was hardly any space left not covered by organisms. The abundance here was dominated by polychaetes (*Polydora ciliate*) and barnacles (*Balanus crenatus*). Unlike on the steel plates, blue mussels could be detected at the steel cylinder only in the juvenile stage. This difference can be attributed to the predation pressure of crabs and starfish, for which the bottom structure was much easier to access than the steel plates suspended in mid-water. They were also favoured by the salinity conditions in the bottom water, where they met their salinity dependent distribution boundaries. Other frequent predators of blue mussels observed in video sequences were 1 - 2 year old species of Baltic cod, which found excellent shelter and prey in and around this structure.

The abundance of *Balanus crenatus* was highest, with about 40 % of all species and 90 % of total biomass. At the surface substrates, this species was replaced by the brackish (*Balanus improvisus*). The scratch samples at the cylinder yielded 15 different taxa.

On all substrates, a total of 41 macro-zoobenthos taxa could be detected after 470 days of exposure. With nine and twelve different species, crustaceans and polychaetes dominated; molluscs and hydrozoans followed with five species each.

The biomass at the deeper levels seemed to have reached an equilibrium after three months, but the increase continued at the surface (Fig. 9). This was due to the growth of *Mytilus edulis*, which reached a biomass of close to 1.9 kg afdw per m² and 95 % of total biomass at the 5 m depth level.

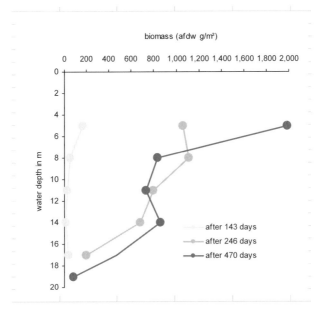

Fig. 9. Development of biomass in different water depths after 143, 246 and 460 days of artificial substrate exposure (from April 2003 to July 2004)

In order to estimate the dependence of colonisation dynamics on the seasonal positioning of the experiments, three new sets of steel tiles were installed in July 2004, and the first of them retrieved and analysed after 177 days in January 2005. The primary exposure period for the first experiment was during summer, while the second set collected species settling in autumn and winter. The differences between the experiments were not significant concerning species diversity. In the first study, 18 taxa were recorded after 143 days; in the second, 20 taxa after 177 days. In both cases, molluscs, polychaetes and crustacean dominated, but while the order of dominance was crustaceans first, then molluscs, and finally polychaetes during summer exposition, the autumn ranking was polychaetes first, then crustaceans, and finally molluscs (Fig. 10).

More significant differences between the experiments could be recorded in relation to macro-fauna abundance and biomass development. The deployment of substrates in a later seasonal phase led to a faster increase in biomass and abundance. The biomass on these substrates reached the one year level of the earlier depositions already after 173 days, and the values at that time were higher by a factor of between 3 and 45, depending on water depth. A comparison is shown in Fig. 11.

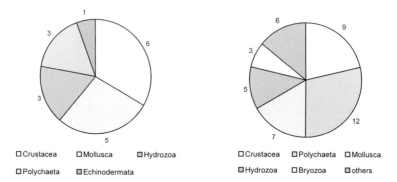

Fig. 10. Succession of macrobenthic diversity on artificial hard substrates (steel plates) after summer exposure from April to August 2003 (left graph) and after winter exposition from July 2004 to January 2005 (right graph). The substrates were retrieved after 143 and 173 days, respectively

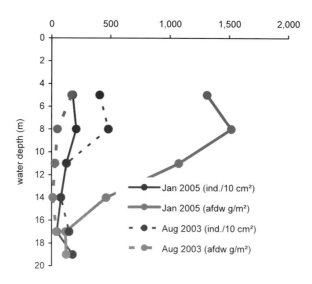

Fig. 11. Abundance and biomass development on substrates with approximate the same exposure time (177 days, Jul. 2004 to Jan. 2005, and 143 days, Apr. 2003 to Aug. 2003), but different seasonal starting dates (July vs. April). Note, that biomass (afdw) is given on a m² base, whereas abundance values are based on 10 cm²

Evidently, the dynamics of colonisation are, at least in the first phase, highly dependent on the seasonal point of deployment of the installation. Figure 12 provides a visual impression on the settling dynamics after retrieval of the substrates from different depths in January 2005.

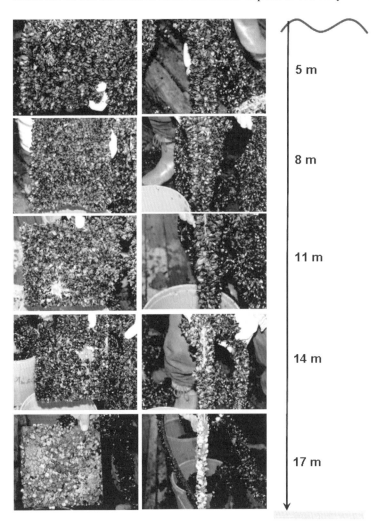

Fig. 12. Epibenthos on settling substrates in different water depths after 177 days of exposure in January 2005. The front and side views of the steel tiles show the increased density of organisms at water levels near the surface

13.3.3 Impact on Sediment Structure and its Living Community

A secondary effect of offshore structures on benthic distribution patterns is exerted by the hydrodynamic changes of turbulence and currents at the base of installations. Figure 13 shows the sediment thickness around the test cylinder after a few months of deposition and an underwater picture of more than 40 cm deep erosion troughs after 246 days.

After this stage, the trench was consolidated by larger shell pieces, and its depth stabilized at 40 cm. The trough then slowly began to fill up with organic debris supplied by flakes of rust with attached organisms from the upper part of the cylinder. Initially, this rich organic load attracted preda-tors (Fig. 14), but later, it created oxygen deficiency, even in the well aerated sediment area. Finally, the organic overload promoted microbial sulphate reduction and the formation of toxic hydrogen sulphide. Figure 14 shows the immediate vicinity of the cylinder after 697 days, with white mats of sulphide oxidising bacteria dominating the picture.

The development of macro-benthic (endo- and epi-benthic) species in the sediment surrounding the steel cylinder was monitored during the entire period of the experiment. In general, the faunistic composition was similar to that of the long term reference station (Stn. 030) nearby. In the beginning, species composition and frequency of appearance were compa-rable to the reference, whereas in a later stage, differences started to develop. After an exposure of 417 days, no significant difference in diver-sity and biomass could detected, while after 697 days the effects of the in-creasing input of biomass from the higher parts of the cylinder began to affect the infauna of the surrounding sediments. Up to that date, the overall effect on the integrated biomass around the cylinder had been positive due to the increase in food supply. Negative trends due to the increased spread of anoxia from the immediate base of the pile were at that time not yet detectable for the wider surrounding area. The quantitative analysis of endo-benthos data was, however, generally hampered by the small number of parallel samples, which could be taken around our comparatively small model substrate, without changing the environment itself. Taken into account the large natural spatial heterogeneity of endo-benthos in such sediments, the results cannot satisfy statistical standards. At this point, the limits of small model substrates have obviously been reached and the necessity for a full size research structure becomes obvious.

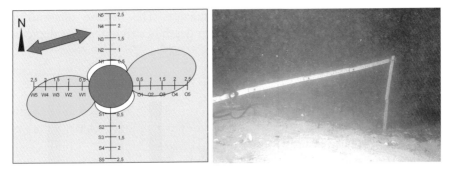

Fig. 13. Direction and spatial extent of sediment erosion (white area) and deposition (grey area) around the test cylinder (dark grey). After 246 days, an erosion depth of more than 40 cm at the cylinder walls could be recorded. The spatial pattern and intensity of the erosion/redeposition process is closely related to main direction of current and speed of current

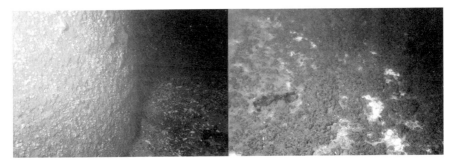

Fig. 14. Succession of biota at the base of the cylinder. After 417 days of exposure (left picture), predators gather at the base, profiting from the first organic material dropping from the upper part of the structure. After 697 days, (right picture) the organic load exceeds the oxygen supply, and the sediment turns anoxic

13.4 Discussion

As a primary effect of the installation of steel structures in the western Baltic waters, a general increase in diversity, abundance and biomass of benthic macro-fauna on the new substrate over time was observed. It is important to note, that the dynamics of colonisation and biomass development are highly dependent on the point in time of the first deployment within the seasonal cycle of larval settlement. The species diversity was dominated by the taxonomic groups of crustaceans, polychetes and molluscs. The total number of species increased during the observation period to 41, with still increasing tendency. The inverse relationship between diversity and biomass with water depth is striking, and is depicted in Fig. 15.

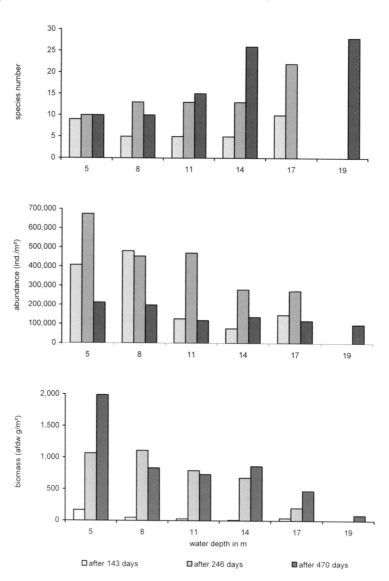

Fig. 15. Temporal development of species diversity, species abundance and biomass during the colonisation experiments at different water depths

Due to missing comparable substrates at the same water depth of a wider surrounding area, the species composition at the model pile is very different from the more endo-benthic species dominated sandy sediments of the Mecklenburg Bight or the Darss Sill (Zettler et al. 2000, 2004). As the vertical extension of the water column coincides with the salinity gradient (Fig. 5), the general decline of species diversity towards poorer brackish conditions (Remane 1955) is mirrored in this development. The increased maturity of the macro-benthos populations at all depths is reflected in the return of the abundance from a steep increase in the first phase to lower numbers, that develops along with the size structure of a mature population (Rumohr et al. 1996). In the context of this development, the biomass in the deeper layers also reached an equilibrium level after 246 days. An exception was the surface population, which increased its biomass through the last sampling period.

The fact that the increase in abundance and biomass is highest at the upper parts of the installations is most probably due to two reasons. The availability of food of good quality and quantity is much higher in the surface mixed layer, where filter feeders can take it out of an abundantly producing pelagic system. The inhabitants of the lower levels of the frame have to content themselves with food that has already left the productive cycle, is less in mass and less nutritive. The feeding process in the vicinity of the sediment may, at the measured current speeds, often be inhibited by large amounts of resuspended mineralic particles which further reduce the nutritive value of the prey. A second factor is the presence of large benthic predators, like starfish, crabs and juvenile cod, which are all restricted to the lower water levels, due to the low saline conditions at the surface. So the observed depth gradient in biomass development of epifauna is easy to explain, and in the first phase of development poses no environmental problems. However, in the more advanced states of fouling at structures, precisely these conditions lead to severe environmental drawbacks, specifically under stratified Baltic conditions.

If, due to the inhibiting surface salinities, predators like starfish, crabs and cod are absent or barely active, but the prey, in this case mainly blue mussels, is not affected, then the typical biological regenerative cycle does not work. There is no biological limit to *Mytlis edulis* growth, so that no biomass equilibrium can develop. At other natural hard substrates, space would sooner or later become the limiting factor, and the growth rate would then adapt. On uncoated steel structures, there is a continuous provision of new substrate whenever the population size and the weight of the epifauna reaches a critical value and disengages the rust flake from the steel surface. This sequence leads to continuous exponential growth and biomass formation on the upper part of steel structures. The biomass is then transferred to the base of these structures as attachment to the large rust flakes. Due to the great weight, this transfer is rapid, and restricted to the immediate surroundings of the piles. In a first phase, the high organic input is appreciated, used and converted by predators, which tend to accumulate at the base (Fig. 14, sketch Fig. 16). In this process, an imbalance between the local supply of organic mater and the availability of oxygen will develop. At the onset of anoxia, microbial metabolism will dominate the decomposition, and oxidants other than oxygen will be used (Fig. 14). As sulphate is present in seawater in large quantities, sulphate reduction very quickly becomes the primary decompositional pathway. The resulting formation of hydrogen sulphide, as a dissolved gas, which is toxic to higher organisms by inhibiting their respiratory chain, initiates a vicious cycle in the enclosed bottom layer which ultimately results in a situation which is already in a critical state in many areas. As a result of increased surface productivity and export of food to the sediments, the macro-fauna biomass has generally increased over the past decades in the southern Baltic (Cederwall and Elmgren 1980, Cederwall et al. 2002, Rumohr et al. 1996) and with it, the demand for oxygen by the sediments (Karlson et al. 2002, Powilleit and Kube 1999, Weigelt and Rumohr 1986). As frequent stratification prohibits additional supply of oxygen by vertical mixing, the benthic systems in these stratified areas have already drifted towards oxygen deficiency over the past decades. If this imbalance is further aggravated by additional input of organic matter, and free hydrogen sulphide starts to spread in the bottom water, large stocks of benthic biomass will be poisoned, and will likewise be subject to anaerobic decomposition, resulting in the emission of even higher amounts of sulphide. If this self powered process gains momentum, a whole marine region can turn into an area devoid of higher animal life. Such a process was observed in the Kiel Bight at the beginning of the 1980s (Weigelt and Rumohr 1986) and in the western Pomeranian Bight in 1994, where just a short term extension of thermal stratification caused a sediment area of several square kilometres to turn sulphidic (Powilleit and Kube 1999). The implications for economy like fisheries and tourism were grave, and the macro-fauna required about half a decade to recover to normal diversity and biomass conditions.

Our model-pile is situated in a better aerated and weaker stratified environment than the areas that are currently disclosed for wind park planning. It is collecting slower growing organic material from far beneath the surface mixed layer, where access for larger predators was possible and it was of a considerably smaller size, that the planned constructions. Nevertheless, the effects became obvious after a short period. Therefore, our observations of the successive negative development in the benthic environment around this test pile led us to assess the construction of large numbers of adjoining full size piles in closer vicinity to the Danish Sound as a potential threat to the benthic habitat in this area.

From the data gathered in our experiment, we calculated the increase of biomass per unit area after the deployment of substrates to range between the factors of 14 and 140. Similar results have been found at Danish installations in the North and Baltic Seas (Birklund and Petersen 2004, Leonard and Pedersen 2004), and in artificial reef systems in Polish waters (Chojnacki 2000). Studies at the Dutch coast have shown that stability of the fouling community is reached after a period of five to six years, which can be extended by storm events and other disturbances (Leewis et al. 2000, Leewis and Hallie 2000).

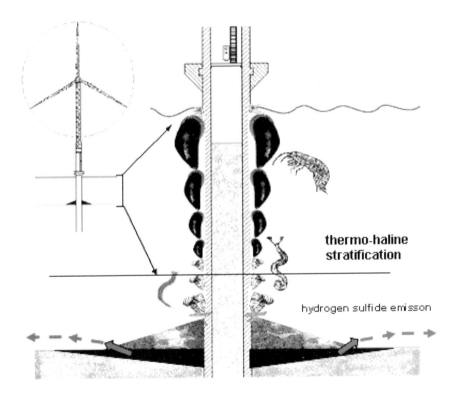

Fig. 16. Sketch of the projected colonisation dynamics on a full-sized wind turbine pile in the western Baltic, based on the results of the present study

The present study shows that in comparison to other areas, biomass increase per unit time in the western Baltic is much higher than elsewhere (Table 1). Biomass decreased with water depth, but was still considerably higher than the initial value of the soft bottom community (see reference station in Table 1).

Taking into consideration the development of biomass on the steel tiles and the bottom cylinder, a steel tube (mono-pile construction) of 2 m diameter in a water depth of 20 m would yield a biomass (wet weight) of 150 kg after 143 days and 1.6 tons after 246 and 470 days. Assuming a pile (tripod) to have a threefold colonisation area in all, and a mid-sized wind farm to have 100 turbines (actual planning in the Baltic calls for between 20 and 400 per farm), the initial annual yield per wind farm comes to about 500 tons. If this is the yearly equilibrium growth rate, balanced by the loss of biomass due to flake detachment, this amount is selectively deposited in small sediment areas, and would certainly have enough impact to start the reaction chain described above and indicated in Fig. 16.

Table 1. Total biomass (wet weight) at the reference station, the pile model (after one year of exposure) and literature values from other pile structures in the North and Baltic Seas

Pile structures	wwt in [g/m^2]
5 m	20,000
8 m	15,000
11 m	11,500
14 m	13,000
17 m	4,000
19 m	2,000
Monitoring (030), 2003-2004, reference station	140
Nysted, Baltic, Pile after 1 year[a]	3,000
Nysted, Baltic, Mast after 6 years[a]	14,500
Horns Rev, North Sea, Pile after 1 year[b]	2,800

a = from Birklund and Petersen (2004). Wet weights were calculated from dry weights by a factor of 2.55
b = from Leonard and Pedersen (2004)

Strategies for mitigation of these problems include the proper distancing of the piles to prohibit the accumulation of single-pile effects to a combined area. Another strategy, which might even be economically sustainable, is to let the mussels grow on removable substrates on the upper part of the piles and harvest them.

Then the succession of adverse processes would be terminated right at the beginning and turned into a positive direction. At the moment, it cannot be predicted how a pile or a set of piles could influence vertical turbulent mixing processes and thus increase the oxygen supply to the sediments. To increase the vertical oxygen transport by pile design or other technical means and hence foster sediment aeration as a countermeasure would certainly lead to local improvements at the hot spot areas. It would, however, initiate a large scale density change of the bottom water by mixing in warm, less saline water. This could in turn prevent the inflowing water from entering the deeper Baltic basins and aerating them.

All these examples show that Baltic problems related to offshore structures are highly diverse from oceanic or North Sea situations, due to a fundamentally difference in physical transport conditions in this sea. Changes in basic transport patterns in one key locality of the delicate transition zone of the western Baltic may propagate through the ecosystem on a basinwide scale. We therefore believe that for a proper assessment of the ecological effects of large wind farms in the stratified waters of the western Baltic, it is necessary to take all of these processes into consideration, study and balance them, and model the resulting impact on a basin scale.

References

Birklund J, Petersen AH (2004) Development of the fouling community on turbine foundations and scour protections in Nysted Offshore Wind Farm, 2003. Energi E2 A/S Report June 2004: 39 pp

Cederwall H, Elmgren R (1980) Biomass increase of benthic macrofauna demonstrates eutrophication of the Baltic Sea. Ophelia 1: pp 287-304

Cederwall H, Diziulis V, Laine A, Osowiecki A, Zettler ML (2002) Eutrophication and related fields: Baltic Proper: benthic conditions. In: Fourth periodic assessment of the state of the marine environment of the Baltic Sea area 1994-1998. Baltic Sea Environm Proc 82B: pp 0-55

Chojnacki JC (2000) Experimental effects of artificial reefs in the southern Baltic (Pomeranian Bay). In: Jensen AC et al. (eds) Artificial Reefs in European Seas. Kluwer Academic Publ, pp 307-317

Feistel R, Nausch G, Matthäus W, Hagen E (2003) Temporal and spatial evolution of Baltic deep water renewal in spring 2003. Oceanologia 45 (4): pp 623-642

Gill AB (2005) Offshore renewable energy: ecological implications of generating electricity in the coastal zone. J Appl Ecol 42:605-615

Karlson K, Rosenberg R, Bonsdorff E (2002) Temporal and spatial large-scale effects of eutrophication and oxygen deficiency on benthic fauna in Scandinavian and Baltic waters – a review. Oceanogr Mar Biol Annu Rev. 40:427-489

Leewis R, Hallie F (2000) An artificial reef experiment off the Dutch coast. In: Jensen AC et al. (eds) Artificial Reefs in European Seas. Kluwer Academic Publ, pp 307-317

Leewis R, van Moorsel G, Waardenburg H (2000) Shipwrecks on the Dutch continental shelf as artificial reefs. In: Jensen AC et al. (eds) Artificial Reefs in European Seas. Kluwer Academic Publ, pp 307-317

Leonard SB, Pedersen J (2004) Hard bottom substrate monitoring Horns Rev Offshore Wind Farm. Annual Status Report 2003. Elsam Engeneering

Powilleit M, Kube J (1997) Effects of severe oxygen depletion on macrobenthos in the Pomeranian Bay (southern Baltic Sea): a case study in a shallow, sublittoral habitat characterised by low species richness. J Sea Res 42:221-234

Remane A (1955) Die Brackwasser-Submergenz und die Umkomposition der Coenosen in Belt- und Ostsee. Kieler Meeresforschung 11: pp 59-73

Rumohr H, Bonsdorff E, Pearson, TH (1996) Zoobenthic succession in Baltic sedimentary habitats. Arch Fish Mar Res 44:179-214

Weigelt M, Rumohr H (1986): Effects of wide-range oxygen depletion on benthic fauna and demersal fish in Kiel Bay 1981-1983. Meeresforschung 31:124-136

Zettler ML, Bönsch R, Gosselck F (2000) Das Makrozoobenthos der Mecklenburger Bucht – rezent und im historischem Vergleich. Meereswissenschaftliche Berichte 42:144 pp

Zettler ML, Röhner M (2004) Verbreitung und Entwicklung des Makrozoobenthos der Ostsee zwischen Fehmarnbelt und Usedom – Daten von 1839 bis 2001. In: Bundesanstalt für Gewässerkunde (ed) Die Biodiversität in Nord- und Ostsee, Vol 3. Report BfG-1421, Koblenz: 175 pp

14 Effect of Electromagnetic Fields on Marine Organisms

Ralf Bochert, Michael L Zettler

14.1 Technical and Physical Background of Magnetic Fields

Artificial magnetic fields are unavoidable features of offshore wind farms in natural geomagnetic field environments. The movement of the wind over the blades makes them rotate and a connected shaft powers a generator to convert the energy into electricity. This electricity is transmitted by cables over long distances. Operating electric currents always produce magnetic fields, which are essentially dipolar in nature, having a north and a south magnetic pole.

The magnetic field lines of a straight current-carrying wire form concentric circles around the wire. The direction of the magnetic field is perpendicular to the wire and defined by the human right-hand rule, where the thumb of the right hand points in the direction of the conventional current and the fingers curl around the wire in the direction of the magnetic field. Direct electric currents (DC) produce static magnetic fields.

The impact of a magnetic field is described by the magnetic flux density (B). It is defined as the force acting per unit length on a wire carrying unit current (I). The magnetic flux density around a very long, straight wire can be calculated as:

$$B = \frac{\mu_0 \times I}{2\pi \times a}$$

B - magnetic flux density (SI-unit Tesla (T))
μ_0 - permeability of a vacuum $\mu_0 = 4\pi * 10^{-7}$ Hm^{-1}
I - current carried by the wire (SI-unit Ampere (A))
a - perpendicular distance from the wire to the point were the flux is being evaluated (m)

The magnetic flux density or magnetic intensity in the environment of a straight wire depends on the value of the electric current, in so far as the magnetic intensity increases with the electric current. In contrast, the magnetic intensity decreases with increasing distance from the wire.

There exists a great variety of feasible technical solutions to the problem of electric power transfer from offshore wind farms. Both variants, alternating current (AC) and direct current (DC) have been used. In addition to single, unipolar power cables systems, bipolar solutions also exist, in which two wires are arranged bipolar in one or two single submarine cables. Each of these variants will produce a different magnetic field. As in alternating current the magnitude and direction of the current varies cyclically, the magnetic field direction changes accordingly. In addition AC-induced magnetic fields lead to the development of electric fields, but technical designs of power cables are able to shield the environment from them. The magnetic intensity of two parallel wires results from the sum of their single fields, with the magnetic field intensity at a given point increasing if the currents are flowing in the same direction and decreasing if they flow in opposite directions.

As mentioned above, the magnetic field intensity of the environment of a single wire system depends on the electric current (I) and the distance (a) from the wire. Electric currents of 850 Ampere (A) and 1,600 A are characteristic of underwater sea cables. From these an artificial magnetic field of about 3.2 Millitesla (mT) is induced near a single wire at 1,600 A. The magnetic intensity decreases to 0.32 mT at a distance of 1 meter (m) and to 0.11 mT in a distance of 4 m. Even at this distance, the artificial magnetic field exceeds the natural geomagnetic field. Geomagnetic field values range from about 0.02 to 0.07 mT, with about 0.05 mT observed in the North and Baltic Seas areas.

14.2   Geomagnetic Field Detection in Marine Organisms

Evidence for orientation in relation to the geomagnetic field is rare in marine animals. Lohmann (1985) and Lohmann et al. (1995) found magnetic orientation of the western Atlantic spiny lobster (*Panulirus argus*). *Panulirus argus* undergoes an annual mass migration. Thousands of lobsters vacate shallow, inshore areas and crawl seaward in single-file, head-to-tail processions. Lines of lobsters within the same geographical area follow nearly identical compass bearings (Lohmann et al. 1995). This navigation based on a magnetic map sense, whereby the lobsters derive positional information from geomagnetic field (Boles and Lohmann 2003) using magnetic material concentrated in the cephalothorax, particularly in tissue associated with the fused thoracic ganglia (Lohmann 1984).

Sea turtles undertake a trans-oceanic migration in which they gradually circle the north Atlantic Ocean. They can distinguish between different earth magnetic field densities, and possess the minimal sensory abilities necessary to approximate their global position using a bicoordinate magnetic map (Lohmann and Lohmann 1996). This guidance system exists even in young loggerhead sea turtles (*Caretta caretta*), in which regional magnetic fields function as navigational markers and elicit changes in swimming direction at crucial geographic boundaries. Sea turtles are able to distinguish magnetic differences below 9 mT (Lohmann et al. 1999 and 2001).

Evidence for geomagnetic field orientation is also found in fish, molluscs and other crustaceans (Gill 2005). Juvenile salmon (*Oncorhynchus tschawytscha*), European silver (migratory) and yellow (stationary) eels (*Anguilla anguilla*) are able to respond to the earth's magnetic field (Karlsson 1985, Tesch et al. 1992). Lohmann and Willows (1987) found this phenomenon in the nudibranch mollusc (*Tritonia diomedea*), and chitons have radulae (tongues) that are covered by ferro-magnetic denticles which enables *Chaetopleura apiculata* to react to variation in ambient magnetic conditions (Ratner 1976). Sandhoppers (*Talitrus saltator*) also orient themselves towards magnetic fields, as has been revealed by experimental studies (Arendse and Kruyswijk 1981).

Barnwell and Brown (1964) found in the mud snail (*Nassarius obsoletus*) a response to a magnetic field only about nine times stronger than the local geomagnetic field.

14.3   Effects of Static Magnetic Field on Biological Systems

Magnetic fields interact directly with magnetically anisotropic or ferromagnetic materials, and with moving charges. They are almost unperturbed by biological tissues (Repacholi and Greenbaum 1999). Static magnetic fields may interact with living systems through magnetic induction (forces on moving ions in solution), magneto-mechanical effects (torques on molecules and ferromagnetic material) and electronic interactions (altering of energy levels and spin orientation of electrons, Repacholi and Greenbaum 1999). For instance, static magnetic fields can alter the early embryonic development in sea urchin embryos from *Lytechinus pictus* and *Strongylocentrotus purpuratus* by delaying the onset of mitosis (Levin and Ernst 1997).

14.4   Long-term Exposure of Marine Benthic Animals to Static Magnetic Fields

Several marine benthic animals could survive exposure to a static magnetic fields of 3.7 mT for several weeks (Table 1), and no differences occur in survival between experimental and control animals (Bochert and Zettler 2004). Mussels (*Mytilus edulis*) could live under this static magnetic field conditions for three month and the determination of gonad index and condition index during the reproductive period in spring revealed no significant differences from the control group (Bochert and Zettler 2004).

Table 1. Test organisms and test conditions for long-term magnetic field exposure experiment.

Test organism	Number of test animals	Number of control animals	Duration of experiment [days]
Young flounder (*Plathichthys flesus*) (Pisces)	18	6	28
Blue mussel (*Mytilus edulis*) (Bivalvia)	60	40	52
North Sea prawn (*Crangon crangon*) (Crustacea, Decapoda)	30	20	49
Glacial relict isopod (*Saduria entomon*) (Crustacea, Isopoda)	24	8	93
Round crab (*Rhithropanopeus harrisii*) (Crustacea, Decapoda)	30	10	57
Sphaeroma hookeri (Crustacea, Isopoda)	30	30	34

14.5   Short-term Exposure of Marine Benthic Animals to Static Magnetic Fields

Short-term reactions of the benthic crustaceans *Crangon crangon, Saduria entomon, Rhithropanopeus harrisii, Asterias rubens* (Echinodermata), *Nereis diversicolor* (Polychaeta) and young flounder *Platichthys flesus* (Pisces) to an artificial static magnetic field were tested in a laboratory study. Magnetic flux density (B) was approx. 2.7 mT. Test animals could decide to leave, to accumulate or to rest in the section of the experimental aquarium subjected to the magnetic field.

The test organisms *Crangon crangon*, *Nereis diversicolor* and *Platichthys flesus* were collected on 19 June 2002 and 18 July 2002 in the western Baltic Sea at an eulitoral station (54°01.562 N, 011°32.541 E) by using a fishing net of 0.5 mm mesh size, 0.5 m wide and in the case of *Nereis diversicolor* by using a fork. *Saduria entomon* was collected by dredging offshore east of Rügen Island (54°42.352 N, 014°20.215 E) on 28 October 2002. *Asterias rubens* was collected by dredging near shore west of Rostock (54°10.630 N, 011°44.549 E) on 15 December 2002. *Rhithropanopeus harrisii* was sampled by hand catching from docks of a small fishing harbour of Rügen (54°18.727 N, 013°40.922 E) on 29 August 2002.

After transport to the laboratory, animals were kept in plastic aquaria 145 x 240 x 150 mm filled with about 10 mm of natural sediment and ambient Baltic Sea water. Animals were fed twice a week with small pieces of fish or *Mytilus edulis*.

All studies were performed in a cooling room at 10°C, at a salinity of 10 psu and a light/dark cycle of 13.5 h/10.5 h.

Investigations were performed with two ring coils, each 300 mm in diameter, arranged parallel with a distance of 200 mm between them (ELWE GmbH, Germany). A direct current power source (DF 3010 10A) generated a magnetic field up to B=2.8 mT. The artificial magnetic flux density was considerably higher than the total geomagnetic field of approx. 49 µT at Rostock (54°10.756 N, 012°04.804 E), with horizontal intensity of about 18 µT and a vertical component of about 46 µT. The magnetic flux density was measured by using a Model Koshava 4-magnetometer (Wuntronic GmbH, Germany). The maximum magnetic flux density was generated at the coil planes; values decreased towards the middle between both coils. A mean magnetic intensity was calculated, and values between the two single coils ranged about ± 8 % around the mean. All measurements were performed at the axial centre of the coil system, where magnetic field (MF) is highly homogenous.

Aquaria 660 x 18.7 x 260 mm were used for the experiment. The aquaria could be divided into two sections equal in space by positioning a movable glass pane. The coil system was positioned on one side of the aquaria and the sediment-filled bottom was arranged centrally, where magnetic force is homogenous (Fig. 1).

The system generated an artificial horizontally directed magnetic field. At the beginning of the experiment, equal numbers of test animals were placed at ambient population densities in each section, and the experiment was started by removing the inserted glass pane after an exposure time of approx. 1.5 hours. At the end of the experiment after 22 hours, the sections were closed again and numbers of animals at each section were counted. Control samples were run at the same time in a separate aquarium without a coil system.

Statistical analysis was performed by the Wilcoxon test. A statistical level of $P < 0.05$ was considered significant.

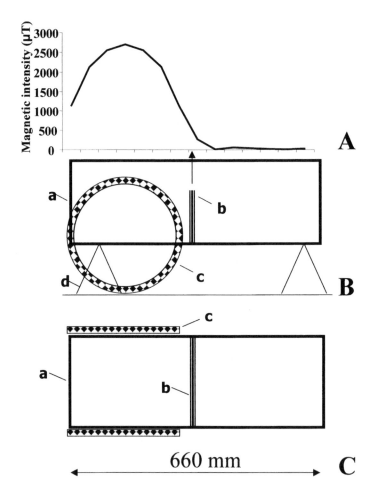

Fig. 1. Study design for magnetic field exposure. a - aquarium, b - movable glass pane, c - coil system, d - table-leg, A - magnetic intensity inside the aquaria along the long side of the aquaria, B - side view of aquaria and coil system, C - top view of aquaria and coil system

The distribution of *Crangon crangon* (six animals per section, 105 individuals per m²) in test aquaria was not different at the end of the experiment, and no significant difference to the control group was evident. The relationship between the sections affected and not affected by the magnetic field, repectively, was well-balanced, at 52 vs. 48 % (n=30). The control group spread was slightly unequal, at 57 and 43 % per section. The highest imbalance measured in both trials of the experiment was 10 to 2 animals, once per test series (Fig. 2).

Saduria entomon (three animals per section, 53 ind./m²) showed a tendency to leave the magnetic field area. Only one third (36 %) of individuals were found in this section at the end of the test series (n=20), whereas the control group was equally distributed. The high standard deviation resulted from a large scattering of single values; differences, at the 5 % level, were not statistically significant (Fig. 2).

Round crab *Rhithropanopeus harrisii* (six animals per section, 105 ind./m²) demonstrated uniform distributions, both for the test runs and for the controls. No differences were recognisable between the two trials. The highest mean values per trial were 54 and 61 % respectively (n=20) (Fig. 2).

The distribution of *Asterias rubens* (five animals per section, 90 ind./m²) remained nearly unchanged at the end of the test series (n=24). Mean values calculated per magnetic field trial peaked at 55 %, and reached 58 % for the control group (Fig. 2).

The tested polychaete *Nereis diversicolor* (six animals per section, 105 ind./m²) showed on average an unchanged distribution in relation to the initial allocation. The animals divided 52 % to 48 % in the test aquaria and 44 % to 56 % in the control group (Fig. 2).

Young flounders *Platichthys flesus*, 1 - 3 cm in length, (three animals per section, 53 ind./m²) showed no significant different distribution when tested with and without the magnetic field. In the magnetic field section, 59 % of the animals in mean were found, whereas without the artificial magnetic field, the mean was 53 % (Fig. 2).

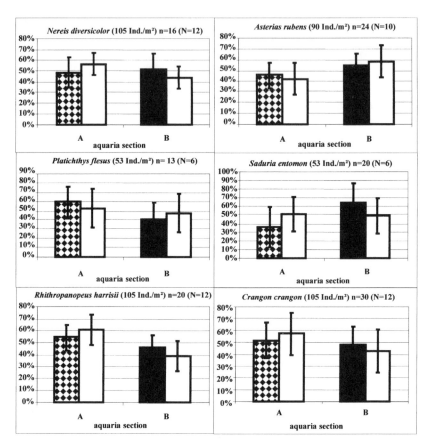

Fig. 2. Short term reaction of *Nereis diversicolor*, *Asterias rubens*, *Crangon crangon*, *Saduria entomon*, *Platichthys flesus* and *Rhithropanopeus harrisii* to a magnetic field. Mean percentage (± standard deviation) distribution of n experimental runs and N number of individuals (number of individuals per m² in brackets) in two aquarium sections. Dotted bar - high magnetic intensity, black bar - low magnetic intensity, white bar - control

14.6   Oxygen Consumption of *Crangon crangon* and *Palaemon squilla*

The oxygen consumption of two North Sea prawns *Crangon crangon* and one *Palaemon squilla* prawn were observed in a closed flow-through system under similar external conditions. Animals were kept three times for three hours in a static magnetic field, in a frequent (50 Hertz) magnetic field of B = 3.2 mT and without a magnetic field.

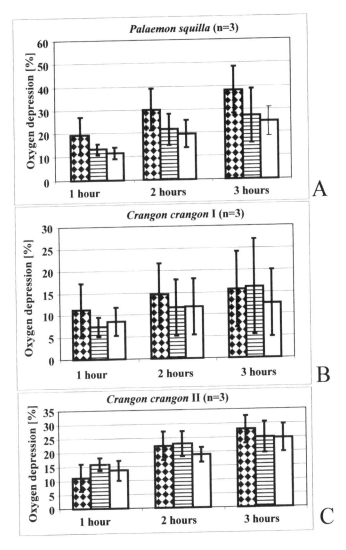

Fig. 3. Mean oxygen consumption (percentage depression) (± standard deviation, n=3) of three decapod crustaceans *Palaemon squilla* (A) and *Crangon crangon* (B, C) during exposure to a B = 3.2 mT static (DC) (dotted bar) and frequent (AC) (lined bar) magnetic field and during control conditions (white bar) after 1 hour and cumulatively after 2 and 3 hours

Oxygen consumption of one *Palaemon squilla* and two *Crangon crangon* showed no significant differences between exposures to static (DC), frequent (AC) magnetic fields, and under control conditions. Cumulated mean oxygen consumption increased in *Palaemon squilla* from 11 - 19 % in the first hour to 24 - 47 % after three hours (Fig. 3A). Mean oxygen consumption was higher during static magnetic field exposure, but differences were not significant. Mean oxygen consumption in two *Crangon crangon* revealed values of 12 - 27 % after three hours and no differences to control were observed (Fig. 3B, C).

14.7 Conclusions

All points on the earth's surface are characterised by the presence of a static geomagnetic field. The amount of total field intensity, which consists of a horizontal and a vertical component, depends on locality, and varies from 20 to 75 μT. However, these low natural values are enough to stimulate reactions in some marine animals of different groups, such as sea turtles (Lohmann and Lohmann 1996a; Lohmann et al. 1999, 2001), fish (Karlsson 1985; Taylor 1986; Tesch et al. 1992), molluscs (Barnwell and Brown 1964; Ratner 1976; Lohmann and Willows 1987) and crustaceans (Arendse and Kruyswijk 1981; Lohmann 1985). Elasmobranch fish are able to sense magnetic fields by their ampullae of Lorenzini (Kalmijn 1982).

In addition to the geomagnetic fields, marine benthic fauna could be subjected to artificial magnetic fields (Gill 2005), which, at a current of I=1,600 A, could produce a magnetic flux density of B=3.2 mT at a distance of 0.1 m, 1.0 mT at a distance of 0.3 m, and even at a distance of 6 m it is in the range of natural geomagnetic field, at about 50 μT. Studies of possible effects of artificial static magnetic field have been carried out in various systematic groups and under various experimental conditions. It has been shown that externally applied magnetic fields could interact with biological systems to produce detectable changes. Often, these findings are very slight differences to control groups, and no clear-cut effects of steady magnetic fields are yet available. In the hydroid *Clava multicornis*, reproduction was faster at a magnetic intensity of 10 and 20 mT than in control and at 40 mT (Karlsen and Aristharkhov 1985). In *Mytilus edulis*, magnetic field action of 5.8, 8 and 80 mT leads to a 20 % decrease in hydration and 15 % decrease in amine nitrogen values (Aristharkhov et al. 1988). Guppies (*Lebistes reticulates*) survived a continuous magnetic treatment of 50 mT for 200 days. In the first generation, but the second generation had an average reduction of spawning rate of 50 % and in the third generation, reproduction was completely inhibited as long as the fish remained within the magnetic field (Brewer 1979).

Our results indicate that all the animals we tested did not react when exposed to an artificial magnetic field. Static magnetic fields of submarine cables seem thus to have no clear influence on orientation, movement and physiology of the tested benthic animals.

Further studies which focus on a long-term approach and different conditions (AC/DC, uni- and bi-polar cables, other species, individual to cellular level etc.) are necessary to confirm fully the harmlessness of power transmission on the marine environments.

References

Arendse, MC and Kruyswijk, CJ (1981) Orientation of *Talitrus saltator* to magnetic fields. Neth J Sea Res 15:23-32

Aristharkhov VM, Arkhipova GV, Pashkova GK (1988) Changes in common mussel biochemical parameters at combined action of hypoxia, temperature and magnetic field. Seria biologisceskaja 2:238-245

Barnwell FH and Brown FA (1964). Responses of planarians and snails. In: Barnothy MF (ed) Biological effects of magnetic field, vol 1. Plenum Press, New York, pp. 263-278

Bochert R and Zettler ML (2004) Long-term exposure of several marine benthic animals to static magnetic fields. Bioelectromagnetics 25:498-502

Boles LC, Lohmann, KJ (2003) True navigation and magnetic maps in spiny lobsters. Nature 421:60-63

Brewer HB (1979) Some preliminary studies of the effects of a static magnetic field on the life cycle of the *Lebistes reticulatus* (Guppy). Biophys J 28:305-314

Gill AB (2005) Offshore renewable energy: ecological implication of generating elecricity in the coastal zone. J Appl Ecol 42:605-615

Kalmijn AJ (1982) Electric and magnetic field detection in elasmobranch fishes. Science 218:916-918

Karlsen AG, Aristharkhov VM (1985) The effect of constant magnetic field on the rate of morphogenesis in a hydroid *Clava multicornis* (Forskal). Zurnal obscej biologii 5:686-690

Karlsson L (1985) Behavioural responses of European silver eels (*Anguilla anguilla*) to earth geomagnetic field. Helgoländer Meeresunt 39:71-81

Levin M and Ernst SG (1997) Applied DC magnetic fields cause alterations in the time of cell divisions and developmental abnormalities in early sea urchin embryos. Bioelectromagnetics 18:255-263

Lohmann KJ (1984) Magnetic remanence in the western Atlantic spiny lobster, *Panulirus argus*. J Exp Biol 113:29-41

Lohmann KJ (1985) Geomagnetic field detection by the western Atlantic spiny lobster, *Panulirus argus*. Mar Behav Physiol 12:1-17

Lohmann KJ and Lohmann CMF (1996) Detection of magnetic field intensity by sea turtles. Nature 380:59-61

Lohmann KJ and Willows AOD (1987) Lunar-modulated geomagnetic orientation by a marine mollusk. Science 235:331-334

Lohmann KJ, Pentcheff ND, Nevitt GA, Stetten GD, Zimmer-Faust RK, Jarrad HE, Boles LC (1995) Magnetic orientation of spiny lobsters in the ocean: experiments with undersea coil systems. J Exp Biol 198:2041-2048

Lohmann KJ, Hester JT, Lohmann CMF (1999) Long-distance navigation in sea turtles. Ethol Ecol Evol 11:1-23

Lohmann KJ, Cain SD, Dodge SA, Lohmann CMF (2001) Regional magnetic fields as navigational markers for sea turtles. Science 294:364-366

Ratner (1976) Kinetic movements in magnetic field of chitons with ferro-magnetic structures. Behav Biol 17:573-578

Repacholi MH and Greenbaum B (1999) Interaction of static and extremely low frequency electric and magnetic fields with living systems: health effects and research needs. Bioelectromagnetics 20:133-160

Taylor PB (1986) Experimental evidence for geomagnetic orientation in juvenile salmon, (*Oncorhynchus schawytscha* Walbaum. J Fish Biol 28:607-623

Tesch FW, Wendt T, Karlsson L (1992) Influence of geomagnetism on the activity and orientation of the eel, *Anguilla anguilla* (L.), as evident from laboratory experiments. Ecol Freshwater Fish 1:52-60

Technical Analyses

15   Installation and Operation of the Research Platform FINO 1 in the North Sea

Gundula Fischer

15.1   Background

The Federal German Government has set the target of doubling the share of renewable energies used by 2010. In relation to the initial year 2000, this means a share of approximately 12.5 % of electric power generation in 2010. After 2010 this expansion is to be continued at a high level, so that by 2050, at least 50 % of our energy supply should be based on renewable energies.

The utilisation of renewable energies, such as offshore wind energy, makes a significant contribution towards environmental protection by reducing emissions of greenhouse gases. Expanding the proportion of wind energy – and of other renewable resources – hence ensures an ecological and sustainable power supply, and does therefore play a vital role in long-term protection of the Earth's ecosystem.

The research project FINO[1] was initiated with a view of determining the effects of such offshore plants on marine flora and fauna. A comprehensive series of measurements are currently being performed on a research platform in the North Sea, involving multidisciplinary investigations into meteorology, oceanography, biology, sedimentary geology and technical aspects. The results are expected to yield findings of great significance for the technical and environmental, but also the economic assessment of offshore wind technology.

Germanischer Lloyd WindEnergie GmbH (GL Wind) has been entrusted with coordinating the design, construction, commissioning and operation of the platform.

However, even with the use of renewable energy sources, there may still be an impact on the environment and nature; any conflicts of interest must be resolved and technological development must be controlled to ensure it is compatible with the environment and nature.

---

[1]   Forschungsplattformen **I**n **N**ord- und **O**stsee (Research Platforms in the North Sea and Baltic)

As part of the FINO project, the possible effects on the marine flora and fauna are to be determined and outlined. These investigations are also intended to permit the advancement and assessment of measures which might then be applied in the extension phases of offshore wind farms to reduce and prevent adverse effects.

Fig. 1. Research platform FINO 1 in the North Sea (Photo: Hero Lang)

15.2   Goals

The objective of the FINO project is to improve the available knowledge on the meteorological and hydrological conditions at sea, and to ascertain the concrete impact that offshore wind turbines may have on the marine flora and fauna. The data obtained from the research platform will also provide the utilities, approval authorities, planners and operators of wind turbines with a profound basis for the determination and assessment of long-term impacts.

Institutes, standardisation bodies and certification organisations will use the results to cross-check and validate the requirements derived from other fields (onshore wind energy and offshore technology). For manufacturers of wind turbines, the findings will lead to designs which are better adapted to offshore conditions.

In keeping with the strategy paper of the Federal Government, the results can serve to promote technical developments and accelerate approval procedures.

Ultimately, through the increase in knowledge in the field of offshore wind energy will enable advances in the development and generation of wind energy at sea, which could also greatly affect the labour market in the future.

15.3   Location in the North Sea

The location of the first research platform FINO 1 in the North Sea is about 45 kilometres north of the island of Borkum, in a water depth of 28 metres (Borkum Reef, coordinates N 54° 0.86' E 6° 35.26'). In the adjacent area, the Federal Maritime and Hydrographic Agency (BSH) has already approved, in November 2001, the construction of the first offshore wind farm. Meanwhile, further offshore wind farm projects have been approved in the North and Baltic Seas.

To push forward the expansion of wind energy in the German North and Baltic Seas, a group of major players in the industry have established an "Offshore Foundation". The goal of this "Foundation of the German Industry for the Utilisation of and Research into Wind Energy at Sea", sponsored by the BMU, is to set up an innovative test field.

The favoured area for this test field, where twelve prototypes of German wind turbine manufacturers are planned for 2007, is in the immediate vicinity (400 metres distance) of FINO 1.

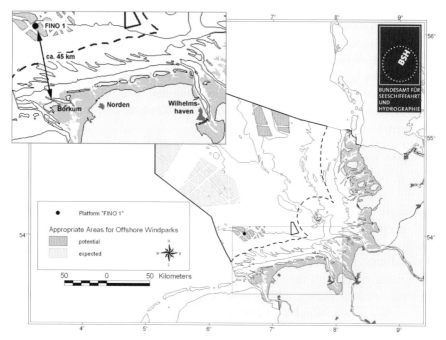

Fig. 2. Location of FINO 1 in the North Sea (Source: BSH)

15.4   Life Cycle

The FINO project started in 2001. For the project's life cycle, the focus of attention was initially on determining a suitable site for the research platform. This was followed by seabed studies, the development of the structure, fabrication and installation at sea.

The seabed studies conducted during October 2001 in the area of the Borkum Reef provided the prerequisite for calculations of the foundation structure for the platform. But also the other environmental conditions such as wind, waves, currents, sea ice, etc. were important for the platform design. The existing "metocean" data were compiled and expertises gathered. All this information, which determines platform design, can be summarised as the "Design Basis", and is displayed in the following table.

The design of the platform began in autumn 2001. With the completion of the tender documents in the spring of 2002, the EU-wide tendering procedure was initiated. It was closed by the submission, and after the valuation of the biddings the tender was accepted in June 2002.

The construction of the platform components started in the summer of 2002. Both the foundation structure and the platform deck were fabricated in concurrent operations at different sites. In the summer of 2003 the components were transported to the location Borkum Reef, and the platform was erected. Operation started in the late summer of 2003. In April 2005 the contracting authority (BMU) assumed ownership of FINO 1.

Table 1. Design Basis (extract) of the location Borkum Reef

Design Parameter	Value	Reference
Water depth	28.0 m	relative to C.D.[a] (= MLWS[b])
Soil conditions	fine sand	0-5 m below sea bed
	medium sand	5-15 m below sea bed
	fine sand	15-32 m below sea bed
Main wind direction	210° - 240°	
Wind speed	49 m/sec	100 year gust (3 sec mean) at 10 m height
Design sea level	+ 5.7 m	relative to C.D.
Design water depth	33.7 m	relative to C.D.
Wind induced current	0.9 m/sec	100 year current/still water level
Tidal current	0.6 m/sec	100 year current/still water level
Significant wave height	7.8 m	100 year sea state
Design wave height	17.0 m	100 year wave
Sea ice	not relevant	
Ice cover	3 cm	on all parts above the sea surface
Water temperature	+ 0.5/+ 20°C	min./max. temperature
Air temperature	- 20/+ 40°C	min./max. temperature

a   C.D. = *Chart Datum* (corresponding to MLWS before 2005)
b   MLWS = *Mean Low Water Springs*

15.5   Structure of the Platform

15.5.1 Foundation

The foundation of FINO 1 consists of four piles, each 38 m long and 1.5 m in diameter. They were driven 30 m into the sea bed and connected to the foundation structure by high-strength concrete, so called grout. The piles are characterised by high rigidity and simultaneously low weight (37 tons each).

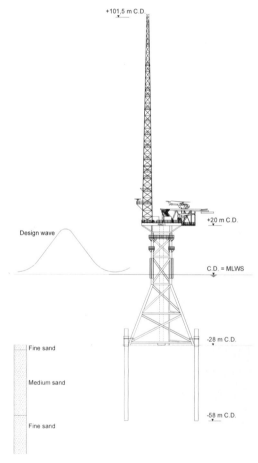

Fig. 3. Research platform FINO 1, view from the south-east

15.5.2 Sub-Structure

The investigation of alternate foundation variants indicated that, for the given water depths, a jacket structure would be the most suitable for the measurement platform. Both the financial and structural results (e.g. low extension) were favourable. To make the berthing process as convenient as possible, a vertical structure was chosen in the docking zone. The welded steel structure has a square ground plan of 26 x 26 m at the sea bed. The height of the jacket reaches 48 m. There are various attachments mounted to the jacket, such as vertical ladders, an intermediate landing at + 10 m chart datum (C.D.), brackets for measurement equipment, a vertical guide-rail for an underwater camera, and a corrosion protection system.

15.5.3 Platform Deck

The platform deck is fixed to the jacket at a height of 20 m above C.D. This height results from the design water level (5.7 m) plus the wave crest (12.3 m) plus a gap (2 m). The size of the welded steel structure is 16 x 16 m. It accommodates five containers, which house measuring equipment, living/working space (including emergency accommodations), a diesel/generator set with batteries and radar equipment, and other equipment.

Fig. 4. Platform deck with helicopter pad and containers (Photo: Hero Lang)

15.5.4 Helicopter Pad

The research platform has been equipped with a helicopter pad. This assures a large time window to access the platform, even in the event of rough weather. The helicopter pad is located 5 m above the platform and equipped with a stairway unit, safety nets all round and navigation lights. Beneath the helipad are the landing and a crane for benthos sampling.

15.5.5 Wind Measurement Mast

The most striking component of FINO 1 is the 80 m high wind measurement mast, reaching a total height of 101 m above C.D. at its top, which is similar to the hub heights of future offshore wind turbines. The steel lattice mast structure is equipped with fold-away booms, where the meteorological sensors are installed. The "met mast" can be climbed by a vertical ladder on the outer side and has resting landings every few meters. Various antennas (radio link, GMS[2], AIS[3], radar) as well as navigation lights are attached to the mast.

15.5.6 Equipment

In addition to the above, some further equipment is worth mentioning.

FINO 1 is equipped with two crane units. The 5-ton offshore crane has an elevating height of 33 m and is normally used for the transport of material. It can also lift containers, the passenger transport cage and, in case of emergency, lower the life raft, even during a power breakdown. The second unit is the "benthos crane" below the helipad, which has a telescopic jib with an outreach of up to 20 m, taking sea bed and water samples.

The electric equipment mainly consists of an energy supply centre with two diesel generators: a 20 kVA generator with 16 kW continuous rating for permanent operation, and a 100 kVA generator with 80 kW continuous rating, used for increased power needs such as crane operation.

Comprehensive safety equipment is available on the platform, including fire protection equipment with CO_2 extinguishing units, optical and acoustic alarm systems, portable fire extinguishers, two life rafts (to be ejected or lowered) as well as several lifebuoys, immersion suits, lifejackets, etc.

---

[2] GMS = Global Messaging Service
[3] AIS = Automatic Identification System

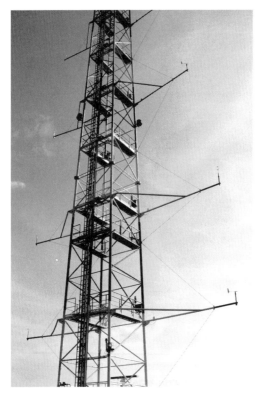

Fig. 5. Wind measurement mast with meteorological sensors (Photo: Deutsches Windenergie Institut, Wilhelmshaven, DEWI)

15.6   Construction and Installation

As already mentioned, both the sub-structure and the platform deck with the helicopter pad were fabricated in concurrent operations at different sites.

The jacket structure was welded at a shipyard area in Bremerhaven. At the same time platform deck and a helicopter pad were manufactured in a hangar in Bremen, which was rented for this particular purpose. The coatings of the deck structure and of the wind measurement mast were also carried out there.

Fig. 6. Transport of the jacket by floating crane (Photo: GL Wind)

After completion of the construction of the sub-structure, the jacket was picked up by a floating crane and transported to the location Borkum Reef, the latter with the aid of several tugboats. There, it was accurately aligned and lowered to its deployment position. To drive the piles, a jack-up platform was positioned exactly at the pile sleeves beside the jacket. The four piles were driven about 30 m into the sea floor and connected to the jacket by grouting.

In the meantime, the platform deck including a helipad, containers with equipment, cranes and even two of the four segments of the wind measurement mast, was loaded onto a pontoon and hauled to the location by tugboats. Here, the floating crane picked up the platform deck and mounted it on the jacket – a demanding task under the prevailing sea conditions. Again the connection between the jacket and deck was grouted.

Finally, the upper two segments of the wind measurement mast were set up on the jack-up platform by a tall crane. All construction and installation activities were supervised by the classification society Germanischer Lloyd.

The main installation work was completed successfully in August 2003. The communication to the mainland by radio link was established, and the scientific institutes carried out the remaining installation works. Since September 2003 FINO 1 reliably provides data from the offshore environment.

Fig. 7. Lifting operation of the platform deck (Photo: Hero Lang)

15.7   Measurements and Investigations

On FINO 1, a comprehensive series of measurements and investigations is being performed to ascertain the environmental conditions (meteorology, oceanography etc.) and the effects on the environment, e.g. benthos, fish, birds and marine mammals. These investigations will also permit the advancement and assessment of measures which can then be applied in the development phases of offshore wind farms, with a view to reducing and preventing adverse effects.

The research on FINO 1 is carried out by various institutes. The following pages will briefly some light on the measurement programme, while the results are described in the respective chapters of this book.

15.7.1 Meteorology

On FINO 1, various meteorological parameters are measured, including:

- wind speed and wind direction at various levels up to + 100 m C.D.,
- air temperature, air humidity, air density,
- global radiation, UV-A irradiation,

- rainfall, and
- lightning.

The meteorological parameters are measured by means of the corresponding sensors, which are mounted at various heights on the measurement mast and at the frame of the platform. Time series are logged for most of the measured parameters.

The results make an important contribution towards improving the available data for the maritime region under investigation. The evaluation and application of the data will enable reduction of the existing risks in the design, construction and operation of offshore wind turbines. As a result, manufacturers and investors will have greater security with regard to a number of aspects of plant construction and assessment of profitability.

In addition, the meteorological data obtained from the platform are used to improve weather forecasting in general.

15.7.2 Oceanography

Within the scope of the project, various oceanographic parameters are measured, such as:

- wave height, period and direction,
- current velocity and direction,
- water level, and
- water temperature, salinity, oxygen content and water pressure.

These data are intended to provide information on the current and wave loads which are to be expected at future offshore wind farms. In addition, the measurement data will supply further indications of the hydrographic conditions prevailing in the German Bight.

15.7.3 Further Technical Measurements

There are further investigations being conducted at FINO 1:

- load impact on the structure,
- acoustic measurements during pile driving, and
- registration of ship traffic.

The load impact on the structure is measured by extensive sensor equipment which records the platform dynamics. The sensor technology includes acceleration sensors as well as strain gages to monitor strain and material fatigue. The data enable the analysis of the interaction of waves, wind forces and the platform structure, and will indicate areas where care should be taken when designing wind turbines.

The acoustic measurements are part of a broader investigation programme, which aims to develop standard procedures for the determination and assessment of noise emissions by offshore wind farms. In cooperation with biologists, recommendations for acoustic emission thresholds for offshore wind farms shall be formulated.

The registration of ship traffic on FINO 1 is regarded as a pilot project to obtain experience in radar-based traffic monitoring. In a selected area around the FINO platform, shipping traffic was registered quantitatively. A more detailed registration in an even larger area is being carried out by Automatic Identification System (AIS).

15.7.4 Biological Investigations

The accompanying ecological research, the BeoFINO[4] project, is aimed at investigating the possible effects of future offshore wind farms on the marine environment (see chapters 9 and 12). On FINO 1, investigations concerning the following topics are being performed:

- marine growth on the underwater structure,
- benthos, fish and planktonic larvae, and
- migratory behaviour of birds (and bats).

For the documentation of the colonisation processes at the platform pile, a camera rail is mounted at the sub-structure of FINO 1, making it possible to take underwater pictures. The underwater images provide information on species composition, abundance, seasonality and the rate of the colonisation of the pile for different species.

A special crane has been installed to take sea bed and water samples for the investigation of benthos, fish and larvae. The telescopic jib with an outreach of up to 20 m enables sampling along a transect at exactly defined distances from the platform's upper rim. The results of these investigations aim to illustrate the possible modification of benthic communities near the piles due to the effect of the artificial hard substrate as well as the development and occurrence of planktonic larvae and fish through the year.

Bird and bat migrations are monitored by various means, such as radar, optical and acoustic methods. Main goals are to assess data on phenology,

---

[4] Ecological research on offshore use of wind energy on research platforms in the North and the Baltic Sea (BeoFINO) funded by the German Federal Ministry for the Environment, Nature Conservation and Nuclear Safety (BMU)

flight altitudes and species composition. These data shall contribute to assess the risk of collision with offshore wind turbines and the resulting impact on bird populations. Methods for minimising bird strikes will be developed, and the knowledge of the migration of bats over the ocean will be improved.

15.8 Data Transfer

For the organisation of the autonomous platform operation as well as for the registration and transmission of the measured and otherwise raised raw data, a computer network has been installed on the platform, to which about 20 clients are connected by radio link telemetry (13 MHz technology) via the platform server. A band width management organises the subdivision of transmission capacity to the individual clients.

The transmission to the mainland is carried out via 32 Mbit/s radio link to the island of Borkum, where data are fed into the fixed network and then transmitted to the German research network. The raw data are accessed by staff of the institutes involved, and then processed accordingly. Upon completion of processing, the data are made available to the public, e.g. on the project's web site: www.fino-offshore.de (results).

15.9 Platform Operation

Since the construction of the platform in 2003, Germanischer Lloyd WindEnergie GmbH (GL Wind) has been entrusted with its operation and maintenance. The operation of FINO 1 is unmanned and watch-free. The platform is monitored centrally from the shore, and is accessible by both ship and helicopter. However, the accessibility by helicopter is beneficial since it is possible during rough weather, and is time-efficient.

Day trips to FINO 1 for maintenance work, on-site measurements, or other field work are coordinated by GL Wind. Furthermore, GL Wind provides the power supply as well as successful operation of the platform equipment, the computer network and the data transfer. Regular inspections of the platform and its equipment are a prerequesite for the success and efficiency of the FINO 1 research.

A project website has been established to make information about the FINO project and the related measurements publicly available. Information on the project as well as results from the measurements may be obtained under www.fino-offshore.de.

Fig. 8. Maintenance work at FINO 1 (Photo: GL Wind)

15.10 Summary and Outlook

The research platform FINO 1 was built in 2003 to investigate the conditions of the offshore environment and to determine the effects of future offshore wind farms on the marine flora and fauna. FINO 1 is the first offshore research platform worldwide, covering investigations of the broad extent described, which has been a challenge in many respects. The location far off the coast and the comprehensive measurement programme have made high demands on the platform and its equipment.

The design and construction were carried out under extraordinary conditions, at a great distance from the shore and with extreme dependence on the weather. The result is the successful construction, operation and delivery of data. FINO 1 can be viewed as a first obvious guiding light for the German offshore wind industry in the North Sea.

FINO 1 has been in operation and has been delivering data very reliably since 2003. This can be attributed to both accurate planning and distinguished adaptation of equipment and instrumentation to offshore conditions. Some results from the FINO activities, especially the biological investigations, are described in this book. Further results of the technical, meteorological and oceanographic measurements are displayed on the project web site www.fino-offshore.de. Meanwhile, these results are being applied by a number of planners of wind farms in the western North Sea. Further scientific work on the basis of FINO data is being carried out in several projects, such as GIGAWIND 2004 and GIGAWIND*plus* 2005, BAGO 2004 and others. Certification bodies such as GL Wind use the results from the wind and wave measurements for a comparison (Argyriadis et al. 2005) with design values as defined in their certification guidelines (Germanischer Lloyd WindEnergie 2005).

The research platform FINO 1 will be of even greater importance as soon as the first activities of the Offshore Foundation concerning the offshore test field start. Also, additional partners will be involved to carry out research on/in the adjacent seafloor. A large number of research tasks remains, e.g. laser-based wind measurements using the newly developed LiDAR[5] instrument, or investigations into scouring and other highly dynamic sediment processes. FINO 1 is a very important step towards the efficient, environmentally-friendly use of offshore wind energy as a sustainable energy supply.

References

GIGAWIND (2004) Bau- und umwelttechnische Aspekte von Offshore Windenergieanlagen. Abschlussbericht 2000-2003, Förderkennzeichen: 0329894A, Universität Hannover, 60 S., Hannover (see also: www.gigawind.de)

GIGAWINDplus (2005) Validierung bautechnischer Bemessungsmethoden für Offshore-Windenergieanlagen anhand der Messdaten der Messplattformen FINO 1 und FINO 2, Jahresbericht 2004, Förderkennzeichen: 0329944, Universität Hannover, Hannover

Lange B, Tautz S (2004) Bestimmung von Wärme- und Impulsfluss in der marinen atmosphärischen Grenzschicht für die Offshore-Windenergienutzung,

---

[5] **L**ight **D**etection **A**nd **R**anging

(BAGO). ForWind Bericht 2004-02, ForWind, University of Oldenburg, Oldenburg (see also: www.forwind.de)

Argyriadis K, Fischer G, Frohböse P, Kindler D, Reher F (2005) Forschungsplattform FINO 1 – einige Messergebnisse. Tagungsband der 4. Tagung „Offshore-WindEnergie" am 14./15. Juni 2005 in Hamburg, Germanischer Lloyd WindEnergie GmbH, Hamburg

Germanischer Lloyd WindEnergie (2005) Guideline for the Certification of Offshore Wind Turbines, Germanischer Lloyd Rules and Guidelines, IV – Industrial Services, Part 2 – Offshore Wind Energy, Hamburg

16 Standard Procedures for the Determination and Assessment of Noise Impact on Sea Life by Offshore Wind Farms

Karl-Heinz Elmer, Wolf-Jürgen Gerasch, Thomas Neumann, Joachim Gabriel, Klaus Betke, Rainer Matuschek, Manfred Schultz - von Glahn

16.1 Introduction

Offshore wind energy is a new technology created by the merging of classical wind energy technology and classical offshore technology. Wind speeds are considerably higher over the sea as compared to onshore sites, but also the cost per installed kW will increase when moving offshore. The rapid development of wind energy use in Germany is accompanied by an increase of the installed power per wind turbine. In the German areas of the North and Baltic Seas, several large offshore wind farms are planned; each with several hundreds turbines of up to 5 MW each.

The Institute for Structural Analysis (ISD) of the University of Hannover, the German Wind Energy Institute (DEWI) in Wilhelmshaven, and the Institute for Technical and Applied Physics (itap) in Oldenburg are partners in a project on: "Standard Procedures for the Determination and Assessment of Noise Impact on Sea Life by Offshore Wind Farms" which is funded by the German Federal Ministry for Environment (BMU).

The aim of this project (CRI, DEWI, itap 2004) is to study the generation, radiation and attenuation of underwater noise, to develop forecasting hydro sound models of offshore wind converters and future noise reduction methods during pile driving, to determine the impact area of offshore wind farms, to allow the formulation of recommendations for acoustic emission thresholds for offshore wind farms in cooperation with biologists, and to develop standard procedures for the determination and assessment of noise emissions.

The operation and in particular the construction of offshore wind converters induce considerable underwater noise emissions. It is assumed that small whales and seals can be affected by noises from machines and vessels, piling and installation of the wind turbines. Piling, in particular using hydraulic hammers creates high frequency noise with considerable sound power levels. Currently, only little knowledge about the effects of different noises to marine life is available. With a view to determining the effects on the marine flora and fauna and structural design aspects, the research platform FINO 1 (Fig. 1) was erected in the North Sea.

Measurements of the underwater noise during construction of offshore research platforms and numerical investigations are used to develop future forecasting hydro sound models of offshore wind converters.

Fig. 1. Research Platform FINO 1

16.2   Physical-technical Principles

There are differences between the treatment of air borne noise and hydro noise. Basic acoustical parameters such as sound pressure, sound velocity, near range, far range, sound pressure levels and mean levels are introduced in the research project with the special focus on underwater noise.

As for other areas of acoustics, in hydro acoustics, frequency distributions of parameters are necessary in most cases, i.e. spectral presentations. In contrast to conventional noise protection, where a main part of the effects can be described within $^1/_3$ octave spectra, in hydro acoustics it is sometimes necessary to go into further detail. This often leads to a situation where levels with different frequency widths are compared directly, without regarding the appropriate conversion factors.

Furthermore the peculiarities of impulsive noise as it occurs during pile driving are described in CRI, DEWI, itap (2004). Besides mean values that are given by the equivalent sound pressure level (L_{eq}), a measure for the single event is necessary to study the biological relevance of one single impact. We use the single event sound pressure level (L_E) for this purpose. As far as the physical definition is concerned, it is identical with the term event sound pressure level (SEL), but refers only to the single event of one "pile driving hit".

16.3   Measurements of Underwater Noise

Investigations on underwater noise of offshore wind turbines by measurements and numerical simulations concern both: the construction noise and the operating noise. But the impact of pile drivers on the piling during construction will result in substantial noise energy propagation within the acoustically shallow water.

Since no offshore wind turbines have been built in German waters so far, construction noise was measured during the erection of two research and wind measurement platforms: GEO and FINO 1. The pile driving techniques were similar to the procedures that are going to be used for erecting offshore wind turbines. Operating noise was measured on a Swedish offshore wind farm.

16.3.1 Measurements of Construction Noise

The foundation of both platforms has been realised by pile driving. The first platform however (company GEO) was erected in the Baltic Sea while the second one FINO 1 has been founded in the North Sea. Therefore also the influence of different pile constructions and different marine environments, such as the ocean bed, could be studied. The immission values were measured in different distances during the whole pile driving procedure in order to get information about the attenuation with the distance and also to see changes within the sound level as the pile penetrates into the ground.

The FINO 1 construction site is shown in Fig. 2. The platform is built as a "jacket foundation", that is, it is "nailed" to the sea floor by means of four piles with about 1.5 m diameter each. The piles are driven into the sea bed almost completely; only a rest of 3 - 4 m remain for fixing the platform. Water depth was about 28 m. About 80 minutes net ramming time was needed for each pile.

Fig. 2. Offshore working platform with crane and pile. On the right side: FINO 1 platform under construction

A useful quantity for describing impulse noise like pile driving strokes is the peak level

$$L_{peak} \text{ in dB} = 20 \log_{10} (| p_{peak} | / p_0,)$$

where p_{peak} is the maximum positive or negative sound pressure observed and p_0 is 1 µPa. A 60 s record of sound pressure is shown in Fig. 3, while a single pile driving blow is displayed in Fig. 4.

Peak levels of 193 dB re1 µPa were observed at a distance of about 400 m from the pile. Quite similar levels were noticed during the pile driving procedure of the GEO platform in the Baltic Sea. The sound pressure was measured at a distance of 300 m and revealed peak levels of about 196 dB re 1 µPa.

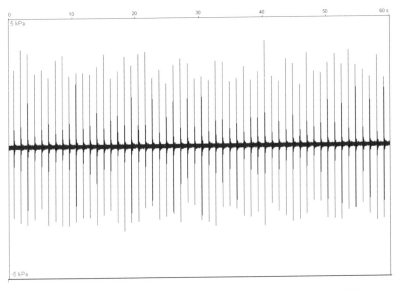

Fig. 3. Pile driving noise at FINO 1. Sound pressure vs. time over a 60 s period at a distance of 400 m

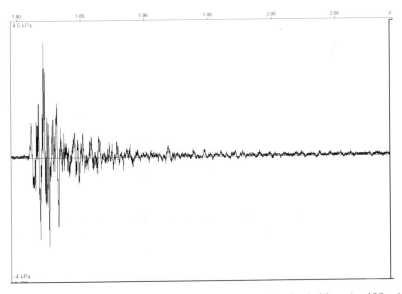

Fig. 4. Pile driving noise at FINO 1. Sound pressure of a single blow (at 400 m)

These values indicate that special care must be taken to avoid harm for animals in the vicinity of the construction site.

For a quantitative study one has to encounter the impact of the single event "one hit" (Fig. 4) together with the cumulative effect of a series of these single events.

In acoustics, time average levels are more common than peak levels. The continuous equivalent sound pressure level (L_{eq}) is an energy average over an arbitrary integration time. The integration time should be chosen short enough in order not to screen the change within the single events, but long enough to show the cumulative impact. A value of ten seconds was found appropriate for impulse rates in the order of one stroke per second.

Such a time series was taken during the building activities for the FINO 1-platform and is shown in Fig 5. The L_{eq} level is clearly connected with the excitation energy that may change during the action, due to changes in the sea bed. For the FINO 1 construction a variation of about 4 - 6 dB during the whole pile driving can be noticed. Though the hammering energy will tend to increase with the penetration depth of the pile, the "loudest" part can be seen in the beginning of the action. This obviously must not be the case for instance for a monopile, that is not buried into the ground and therefore has a constant contact area between pile and water.

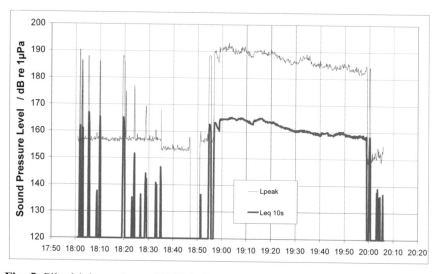

Fig. 5. Pile driving noise at FINO 1: Underwater sound pressure levels at 400 m distance. (Main driving section from 18:55 to 20:00)

L_{eq} and L_{peak} for the whole pile driving process are shown in Fig. 5. The level decrease of 4 - 6 dB towards the end of the process can be explained by the decrease of the radiating pile area while the pile is driven into the ground. Smaller level variations are most likely due to a variation of necessary excitation energy, which results from different layers in the sea bed.

The L_{eq} is useful to display an overview of the whole working process. It is evident, however, that the L_{eq} is affected by the impulse rate (strokes per minute) of the pile driver, working pauses will cause level drops, the depth of which depend on integration time, and so on. Hence for comparing different pile drivers, for evaluating noise reduction techniques and for evaluation the effect on animal hearing, the single event sound pressure level (L_E) (sometimes also abbreviated SEL) is better suitable, which is basically an L_{eq} normalized to one second. L_{eq} and L_E can be converted into each other:

$$L_E = L_{eq} - 10\log\frac{nT_0}{T},$$

where n is the number of impulses in the time interval T and T_0 is 1 s.

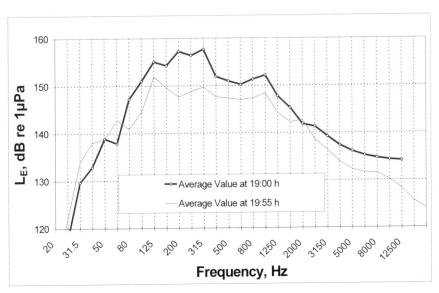

Fig. 6. Pile driving noise at FINO 1; $^1/_3$ octave band spectra measured at 400 m distance

So far only wideband sound levels were discussed, but of course pile driving noise causes a specific spectral distribution.

Figure 6 shows $^1/_3$ octave spectra of L_E. The spectra have a wide maximum between 100 Hz and 400 Hz.

16.3.2 Measurement of Turbine Operating Noise

Vibration of the turbine's gear box and generator is guided downwards and radiated as sound from the tower wall (Fig. 7). Sound radiation by surface waves is difficult to compute and to predict, in particular for complicated boundary conditions. Hence, measurements on an already existing offshore wind turbine were made.

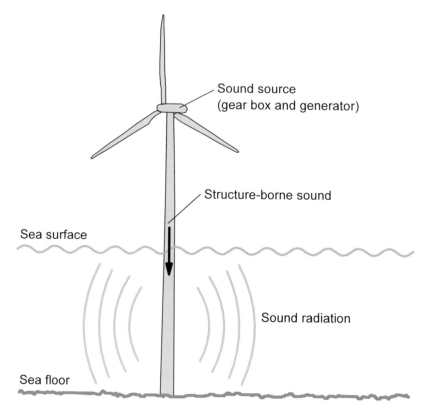

Fig. 7. Mechanism of underwater noise generation by an offshore wind turbine

The measurement setup is outlined in Fig. 8. Since access to the turbine is only possible at low wind speeds, an automatic recording was made over a one month period.

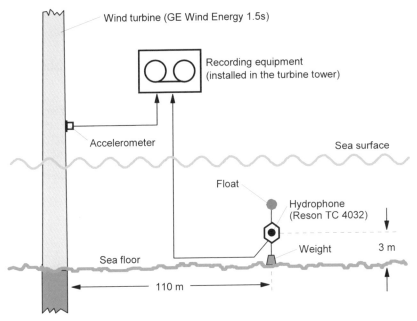

Fig. 8. Measurement setup for monitoring underwater noise induced by an offshore wind turbine (water depth was about 10 m)

At every full hour, 20 minutes of underwater sound and tower wall vibration were recorded to hard disk. The accelerometer position – approx. 10 m above sea level and perpendicular to the wall – was chosen after preliminary measurements with several sensor positions above and below sea level. Wind and electric power values were taken from the turbine's routine log files.

Some acoustic spectra are shown in Fig. 9. At low wind speeds, the generator runs at about 1,100 rpm, but rises rapidly to the nominal value of 1,800 rpm, which is reached at 700 kW. Turbine rated power is 1,500 kW. Hence there are mainly two acoustic spectra (caused by two different sets of tooth mesh frequencies), one for low wind speeds, and one for moderate and strong wind.

The sound levels found here will certainly not cause damage to the hearing organ of marine animals, but might affect their behavior in the vicinity of a turbine. However, somewhat higher tower vibration levels than for this turbine type have been measured onshore on several 2 to 2.5 MW turbines. If set up offshore, these turbine models are likely to produce higher underwater noise levels than those of Fig. 9.

On the other hand, the larger the turbine, the lower the tooth mesh frequencies, radiation efficiency of surface wave declines towards low frequencies, while hearing thresholds increase.

At present, it is not clear if the underwater noise from offshore wind turbine will influence the behaviour of marine animals.

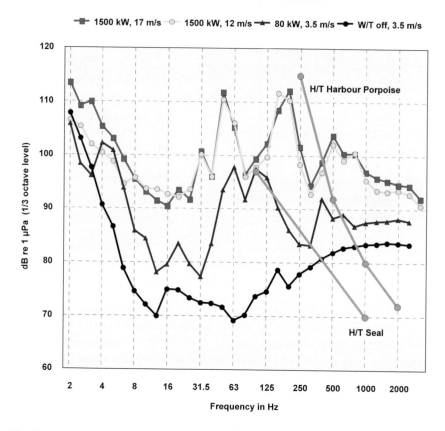

Fig. 9. Underwater sound pressure levels ($^1/_3$ octave spectra) recorded at 110 m distance from the turbine for different turbine states. Wind speeds refer to hub height (nacelle anemometer). Low frequency parts of hearing thresholds for two marine mammals are shown for comparison

16.4   Acoustic Noise Prediction

To be able to assess the noise emission of a wind farm even before its realisation and to estimate its effect on marine animals, it is necessary to have prediction models by means of which a noise immission in the water body can be determined from an existing source level measured at a reference point in the wind turbine. A focal point in such a model is the quantitative analysis of the transition between tower and water, and the transmission of oscillations in the tower itself. It is not possible within the framework of this project to develop a transmission model to such an extent that the results are applicable and reproducible for any type of wind turbine. The results obtained so far, however, have shown that in principle such a model will work.

Two fundamentally different approaches to developing a prediction model are taken. Besides extensive numerical calculations, in which the turbine is treated as a numerical finite-element model, also a simple heuristic model based on tower oscillations and noise emission measured, is proposed. Both methods have advantages and disadvantages: The finite-element modelling and numerical simulations are only reliable if the system tower/water body/sea bottom can be described correctly and verifiably. The empirical model using transfer functions is only useful if the system measured and the system predicted have similar emission characteristics.

16.4.1 Numerical Simulation of Underwater Noise

Monopile foundations represent the most commonly used solution in conventional offshore industries. The impact of pile drivers on the piling results in substantial noise energy propagation within the acoustically shallow water. This construction noise is most important to marine life. The piles are driven into the sea ground by means of a vibrating or piling hammer. Piles are also used to fix tripod and jacket foundations. As an example of this the radiated underwater noise of the FINO 1 platform after Fig. 10 was simulated during the pile driving.

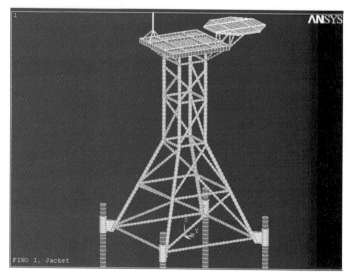

Fig. 10. FINO 1 jacket foundation with piles

Numerical simulations of pile driving, radiation of underwater noise and the propagation of noise are done based on the symmetrical Finite-Element model depicted in Fig. 11 and using the FE-program ANSYS.

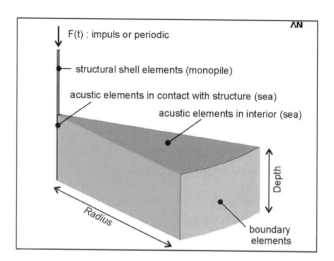

Fig. 11. System of monopile and water with shell and acoustic elements

Figure 12 shows the system of the FINO 1 platform with pile and pile driver. The pile of length l = 37.5 m has stepped cross sections and thickness of the wall between 40 and 18 mm. These sections cause reflections of the impact wave beside the reflections at both ends of the tube. The response of the pile system in the shallow water of 28 m to the impact of the pile driver is axially symmetrical.

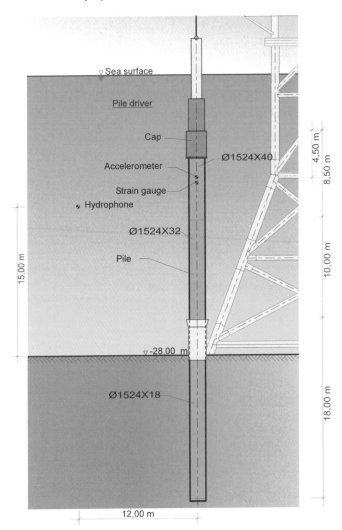

Fig. 12. Foundation of the FINO 1 platform with the hydraulic hammer driving the pile into the ground

In order to obtain reliable results from numerical simulations of the complex mechanism of transient dynamic noise generation and radiation of noise, it is necessary to know the amount and the time function of the impact force as the driving force of the system.

The characteristic number of a pile driver is the maximum impact energy of the hammer. The piles of the FINO 1 foundation with a diameter of 1.50 m were driven into the ground by a hydraulic hammer IHC 280 with nominal energy of 280 kNm. To get the real energy of an impact, accelerations and strain rates of a pile were measured near the driving point at the top of the tube.

Numerical FE simulations yield hydrodynamic pressure at a distance of 12 m and 13 m depth of more than 22000 Pa (Fig. 13) and a typical peak sound pressure level of L_{peak} = 207 dB re 1 µPa.

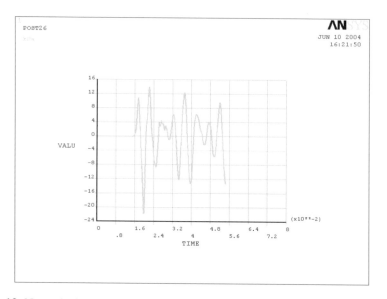

Fig. 13. Numerical underwater sound pressure at 12 m distance of 22,000 Pa. The peak sound pressure level yields L_{peak} = 206.8 dB with respect to 1 µPa

This is in good agreement to measured results of the underwater noise peak level during pile driving of 205.8 dB re 1 µPa in Fig. 14 although the considered frequency range of the numerical model is limited to frequencies below 400 Hz according to time step and element size.

Fig. 14 also shows in the middle of the time function peak values from bumping effects of the driven pile and the pile sleeve of the FINO 1 foundation in Fig. 12.

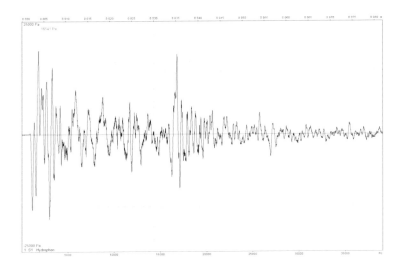

Fig. 14. Measured sound pressure at 12 m distance with a peak sound pressure level of Lpeak = 205.8 dB re 1 µPa

16.4.2 Prediction of Wind Farm Operating Noise

From the levels measured from a single turbine (Fig. 9) and by applying an appropriate sound propagation model, the sound levels produced by a whole wind farm can be predicted. A simple common used model, which however is appropriate for frequencies from 100 - 2,000 Hz and distances up to a few km, is a level decrease of 4.5 dB per distance doubling. For larger distances and for very shallow waters (< 15 m), more sophisticated propagation formulas are required (Thiele 2002). An example for a sound level distribution computed with the simple model is shown in Fig. 15.

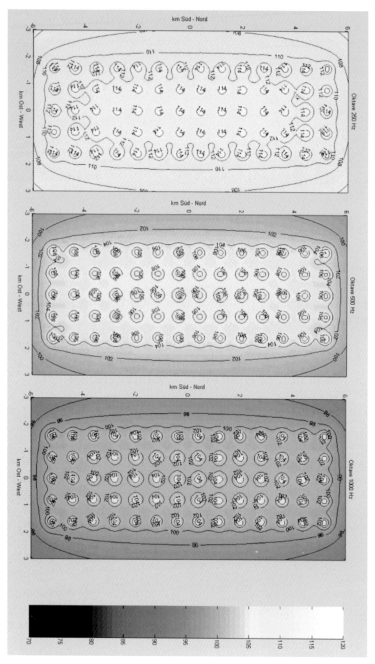

Fig. 15. Underwater noise levels of a hypothetical wind farm of 5 x 14 = 70 turbines spaced 800 m. Levels in in dB re 1 µPa for octave bands 250 Hz, 500 Hz
und 1,000 Hz. The maps cover an area of 6 km x 12 km

16.4.3 Prediction of Turbine Operating Noise with Transfer Functions

It is desirable to predict underwater sound levels of wind turbine types that have not yet been realised offshore, but do exist as prototypes onshore. A semi-empirical approach is the use of "transfer functions". As shown in Fig. 7, the tower vibration is radiated as sound into the water. If the transfer function between vibration and sound for a given foundation type is known, underwater noise levels can be estimated from vibration levels measured onshore.

The principle of deriving a transfer function is shown in Fig. 8. A transfer function measured on a 1.5 MW offshore turbine with monopile foundation is shown in Fig. 16. With a sound propagation model, noise levels can be computed for arbitrary distances.

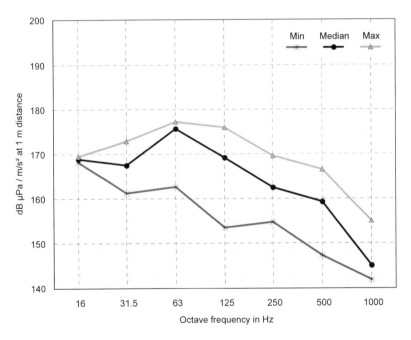

Fig. 16. Amplitude transfer function of underwater sound level vs. tower wall vibration, normalized to 1 m distance. Example: An acceleration of 1 m/s² at 125 Hz produces an average sound pressure level of 170 dB re 1 µPa and a maximum value of approx. 176 dB

16.5   Biological Relevance

The determination of biologically relevant parameters can only be realised in dialogue with expert biologists. On the part of the physicists and engineers, proposals were made for a comprehensive and reproducible description of the hydro sound immission from offshore wind turbines. The selection of the relevant data and the biological evaluation has to be performed by biologists, but should be done in dialogue with the physical/engineering side. The main focus of the physical and biological investigation is construction noise because of its extremely high sound levels

For the approval procedure for offshore wind farms, biologists have established limit values for hydro sound immissions during the construction phase, which, however, are not yet sufficiently backed up scientifically. The working hypothesis for a Temporary Threshold Shift (TTS) for harbour porpoises is a start to put the immission protection during the construction phase in concrete terms. Until detailed scientific findings are available, the experts involved in the project will assume a single event sound pressure level (L_E) of 160 dB (re 1 µPa) at a distance of 750 m as the starting point for an assessment.

A proposal has been made to form a "Working Group for Noise Immission from Offshore Wind Turbines" as a forum for the necessary further expert discussions, based on the model of a similar working group on shore. This group of experts should meet regularly to assess the knowledge gained from physical/technical and biological investigations, evaluate data from accompanying investigations and make scientifically backed up proposals for the approval procedure.

16.6   Standards for the Assessment of Acoustic Emissions of Offshore Wind Farms

In onshore wind energy approval practice, standardised verification and evaluation procedures have been established which have proved to be useful in order to avoid harmful influences on the environment. The Federal Immission Control Act ("Bundes-Immissionsschutzgesetz" – BImSchG) in connection with the Technical Instructions for the protection against noise ("Technische Anleitung zum Schutz gegen Lärm" – TA Lärm) are the basic regulations ensuring the protection of the human environment against noise. An essential part of these regulations are standardised evaluation procedures with a scale providing criteria for the identification of harmful environmental impacts. They are based on comprehensive fundamental research into the human hearing curve, psycho acoustics, propagation of air-borne noise and on measurement and prediction methods specified in great detail.

For the installation and operation of wind turbines special supplementary regulations were established, taking into account the specific characteristics of these acoustic sources (extreme height of source, dependence on wind speed, variable rotor speed, etc.). The most important of the regulations are the "Technische Richtlinie für Windenergieanlagen (FGW-Richtlinie)", the IEA Recommendation on the Measurement of Noise Immission in the Environment of Wind Turbines and the recommendations of the German working group "Wind Turbine Noise".

For the offshore area, the paragraph 15.4.3 of the Standards for Environmental Impact Assessment ("Standarduntersuchungskonzept StUK") of the Federal Maritime and Hydrographic Agency (BSH) serves as a guideline for dealing with hydro sound immission in the approval procedures for offshore wind turbines. However, the approval practice for offshore wind turbines is still far away from the objective evaluation criteria developed for onshore wind energy. There are no limit values for admissible noise immission and no standardised verification procedures.

Limit values for noise impact on the marine environment can only be established in co-operation with biologists and on the basis of relevant research studies (see 16.5 "Acoustic Noise Prediction"). First, however, relevant specific parameters and the standardisation of the measurement, evaluation and presentation methods must be defined, because a limit value x does not make sense when the values of parameters such as L_{eq}, L_{peak} and L_E differ largely and the details of measurements and evaluations are more or less arbitrary.

Standardised procedures for measurement, evaluation and documentation for dealing with the noise immission of offshore wind turbines are necessary. Because of the limited practical experience, however, this complex subject cannot be concluded at this stage. In fact, we are at the beginning of a development process. By means of the procedures described it will be possible to compile a data base of standardised data for example from accompanying investigations on offshore wind farm projects. This data base can be used to check the practicability of the standardisation procedures and fill in the existing gaps in knowledge. It is to be expected that an adjustment of the standards to new findings and requirements will take place in due time.

The measuring and prognosis procedures should be used as suitable components to handle the acoustic noise of offshore wind turbines during erection and operation. Therefore the following requirements should be fulfilled:

- Derivation of biologically relevant parameters,
- reproducibility of the results,
- practicability for monitoring measurements, and
- compatibility of prognosis- and measuring procedures.

16.6.1 Prognosis Procedure

The reliability of the prognosis procedures for the determination of acoustic noise in the building and operating phase depends on the correct description of the inputs of the wind energy converter (WEC), in particular

- WEC parameter such as rated output, rotor diameter, etc.,
- type of the foundation, material, pile depth, etc.,
- effective pile driving and/or vibration energy,
- period length of building process and blow or vibrator frequency, and
- depth of water at the site.

16.6.2 Measurements of the Hydro Acoustic Background

Intensity and spectral composition of marine background noises changes only relatively slow, so that a high temporal resolution is not necessary, even with occasional navigation in the proximity of the hydrophone position. Due to acoustical interference of the ship noise with the measurement a stand alone monitoring system should be preferred.

Items to be documented are:

- Time, date and duration of measurements,
- site, instrumentation, weather conditions (high and low wind speed),
- measuring position in water (preferred 3 to 5 m, above half water depth),
- mean, minimum and maximum values of the equivalent continuous sound pressure level (L_{eq}), and
- $^1/_3$ octave spectra of mean, minimum and maximum values.

16.6.3 Measurements in the Operating Phase

With these measurements the proof is to be led that the established off-shore wind farm keeps the prognosis values. Since so far only a few experiences are present for the measurement of the sound radiation of off-shore WEC, the following statements and suggestions for a measuring procedure are subject to further development.

The plants of future offshore wind parks will be established to each other in large distance; rasters from 800 m to over 1,000 m are usual. It is therefore suggested to measure source levels at individual WECs of the wind farm in a random sampling way. The noise immission of the total wind park can then be extrapolated from these individual measurements.

The entire power range of the WEC shall be regarded, from the start up with low revolution speed of the rotor up to the rated output (= maximum power). Rated power output is correlated to rough sea. For this reason a measurement from the ship is not practicable, in addition an interference with ship noise could falsify the measurements or make them more difficult. Therefore again a stand alone monitoring system should be preferred.

For an offline analysis time signals shall be recorded. For example this concerns the allocation of the measured acoustic noise to particular WEC components by narrow band analysis. During the measurement therefore the WEC data, such as the number of revolutions and electrical power output, should be logged with sufficient time resolution.

Items to be documented are:

- Time date and duration of measurements at 3 arbitrarily chosen WEC,
- site and object, instrumentation, weather conditions, etc.,
- parameters of the WEC such as power output, revolution speed, etc.,
- measuring position in water (preferred 3 to 5 m, above half water depth),
- time signals of the hydrophone for three power ranges of the WEC ("low", "medium" and "maximum"), and
- $^1/_3$ octave and narrow band spectra for the three power ranges.

16.6.4 Construction Phase

Measurements have to be performed outside the near field. A distance in the range of 200 m to 1,000 m is recommended. The hydrophone position should be below half depth but at least 5 m above sea bottom. Broadband recording of hydro sound has to cover all times of relevant immissions.

Equivalent continuous sound pressure levels (Leq) and peak values of the sound pressure level (L_{peak}) have to be measured. The signal of the hydrophone has to be recorded for off-line evaluations (e.g. spectral analyses).

Measurements should give information about duration and amplitude of the acoustic noise immission during the whole construction phase. Therefore a time plot of the equivalent continuous sound pressure level (L_{eq}) and peak values of the sound pressure level (L_{peak}) is required.

Relevant phases for a detailed analysis, e.g. the evaluation of spectra, are times of typical and extreme acoustic noise immission. For each relevant time period third octave spectra of the single event sound pressure level L_E shall be evaluated. In cases of tonal noise characteristics (e.g. vibrators) narrow band analyses are required.

Items to be documented are:

- Time date and duration of measurements,
- site and object, instrumentation, weather conditions, etc.,
- time series of the equivalent continuous sound pressure level (L_{eq}) and peak values of the sound pressure level (L_{peak}) for the whole construction time,
- the equivalent continuous sound pressure level (L_{eq}) and peak values (L_{peak}) for each relevant time period,
- third octave spectra of the single event sound pressure level L_E for each relevant time period,
- plots of the sound Pressure of a single impulse and a series of impulses, and
- duration of the acoustic noise immission and the frequency of the impulses.

16.7 Summary

In the German North and Baltic Seas, claims of wind farms are planned with several hundreds turbines of up to 5 MW each. Furthermore, several research platforms are installed in the sea to determine meteorological aspects, structural design aspects and possible effects of future offshore wind turbines on fish and diving mammals.

Both operation and construction of offshore wind turbines induce underwater noise. The turbine under operation causes only moderate sound levels, but it is a permanent sound source and may affect marine animal behaviour. During construction, the use of pile drivers in particular results in very high levels, which may cause permanent or temporary damage to animal hearing systems.

In onshore wind energy, the noise immission from wind turbines is an important issue under the law concerning the respective rights of neighbours and consequently has been investigated intensively in the past years. In co-operation with wind energy experts the building authorities have developed and introduced standards and regulations which have influenced the design of the wind turbines and vice versa, and so have led to considerable improvements in the noise emission of modern wind turbines. Standardised procedures for obtaining noise certificates in order to ensure immission control have been established.

The present research project aims at providing a theoretical foundation to enable a comparable assessment of noise immission from the installation and operation of offshore wind farms. The following aspects were dealt with:

1. The physical-technical principles for describing structure-borne noise and fluid noise were summarised. In this context also a selection of relevant parameters was proposed upon which the investigations should concentrate.
2. Extensive measurements were carried out in the areas of "acoustic background", "noise during construction" and "noise during operation". As a result of these measurements, the previously poor data situation in the area of offshore noise has improved significantly. Furthermore the experience gained during the measurements could be used to develop proposals for standards for the realisation of such measurements.
3. The prediction of acoustic noise is an important instrument for an early assessment of the noise impact on the marine environment caused by offshore wind farm projects. During this project, first predictions of this

kind were carried out using complex finite-element calculation as well as an empirical approach. For these calculations not only the transition of the turbine structure into the water, but also the sound propagation in the water is important. The strong dependence of sound propagation on environmental influences and the complex interactions in sound immission make it necessary to carry out further investigations especially in the range of predictions in order to achieve the goal of a standardised procedure.

4. As far as the biological importance of offshore noise is concerned, there is increasing evidence that it is mainly the noise caused during the construction phase that is disturbing. The group of experts brought together for this project was able to establish for the first time a working hypothesis on a tolerable limit value. Because of insufficient biological information, however, it was not yet possible to come to a final conclusion in this matter.

However, because of the large number of different projects, the as yet unknown methods of installation, and also because of the strong dependence on the environment of the phenomena investigated, a number of questions especially in the range of prediction could not be answered finally. The thresholds for acoustic noise immission that have been suggested tentatively by the biological-technical experts involved, probably can only be achieved when adopting means to reduce noise (in case of using pulse-driving methods) or by using alternative noise-optimised methods for pile-driving. The investigations of noise-reducing measures as well as the consolidation of the data base and the optimisation of the prediction methods are the objects of a two-year extension of the research project.

References

CRI, DEWI, itap (2004) Standardverfahren zur Ermittlung und Bewertung der Belastung der Meeresumwelt durch die Schallimmission von Offshore-Windenergieanlagen. BMU-Projekt (FKZ 0327528A). Bundesministerium für Umwelt, Naturschutz und Reaktorsicherheit, Berlin

Bethke K Schultz-von Glahn M, Matuschek R (2004) Underwater noise emissions from offshore wind turbines. DAGA'04 (30. Deutsche Jahrestagung für Akustik). Deutsche Gesellschaft für Akustik e.V. (DEGA)

Elmer K-H, Gerasch W-J (2004) Numerical simulation and measurement of underwater noise during pile driving of offshore wind converters; International Symposium on Numerical Simulation of Environmental Problems, November 22-23, Okayama University, Japan

Neumann T, Gabriel J, Gerasch W-J. Elmer K-H, Schultz-von Glahn M, Bethke K (2005) Standards for the assessment of acoustic emissions of offshore wind farms. DEWI-Magazin No 26, Februar 2005

Thiele R (2002) Mitteilung beim Fachgespräch zum UBA-Projekt „Vermeidung und Verminderung von Belastungen der Meeresumwelt durch Offshore-Windenergieanlagen im küstenfernen Bereich der Nord- und Ostsee" am 15.01.2002

17 Collisions of Ships with Offshore Wind Turbines: Calculation and Risk Evaluation

Florian Biehl, Eike Lehmann

17.1 Introduction

Collisions of ships with offshore wind energy turbines (OWTs) constitute a considerable threat to the environment. It must be considered that in a collision incident, parts of the ship's structure will be damaged. Leakage of operating supply or cargo (e. g. oil or chemicals) is possible. In a worst case scenario the ship could break apart and sink.

The research project referred to in this paper undertook a numerical evaluation of several collision scenarios between different ship types and three exemplary types of foundation structure. The resulting conclusions were supposed to lead to an evaluative scheme to determine the mechanical properties of OWT foundation structures concerning their crashworthiness and their ability to conserve hull integrity in ship collisions. These guidelines shall be used in the process of the approval of OWTs.

A stochastic analysis of the probability of collisions was not the goal of the project, although it is necessary to link both an analysis concerning the probability of a collision and a consequence analysis, to determine the risk.

In an analysis done by the Federal Environmental Agency (UBA) on preventive action in the event of a failure in offshore wind farms, a single hull oil-tanker of 160,000 tdw was proposed to be the design ship in the accidental limit state (ALS). Also, a damage of three cargo tanks was estimated as likely, which assumed 54,400 tons of spilled oil as the basis for calculating necessary preventive action (Kremser 2004).

If the mechanical performance of OWTs in case of collision is known, the probability of environmental damage can be estimated, depending on the particular conditions. This leads to a more favourable evaluation of environmental risk than has hitherto been possible. This especially takes into account the increase in passive safety measures against collisions. To provide maximum safety, provisions for active collision and fault event safety, such as redundant navigation and control systems, a ban on navigation for certain kinds of ships, crew training, traffic management systems, wind farm monitoring, tug boats for emergencies, etc.) must be considered in order to prevent collisions and emergency situations before they occur.

17.2   Technical Bases and Numerical Modelling

17.2.1 Collision of Ships

The aspect of collision safety is mostly treated in connection with the construction of tankers. For this type of vessels, there is an international binding agreement (MARPOL 73/78 Annex I, Directive 13F), which determines the minimum dimensions of double bottoms and double hulls. According to the design specifications of Germanischer Lloyd (2002), an extra class index (COLL) can be achieved, if there is calculatory evidence for heightened safety in collisions.

The standard of knowledge and the methods of simulating collision and grounding events were enhanced by scientific projects, which were initiated by the spectacular tanker wrecks of "Exxon Valdez" and "Braer" and set forth e.g. in connection with the construction of the Great-Belt-Crossing. In these projects, empirical, analytical, and numerical methods were applied and many experiments were executed. Several experiments and analyses are described in the dissertation by Zhang (1999), which also features an extensive reading list on the field of collision analysis.

With the aid and the enhancement of these methods and findings, this institute carried out two projects between 1995 and 1999 which dealt with the safety of double hull tankers in instances of collision and grounding (Kulzep and Peschmann 1998 and 1999). Apart from this, there is a worldwide interest in the limitation of risks of collision. An overview of the current situation can be found in the technical literature, especially in the ICCGS-Conference Proceedings[1].

17.2.2 Foundation Structures of Offshore Wind Energy Turbines

The OWTs considered in this study are designed with steel pile foundations[2]. Wiemann et al. (2002) give a survey of explorations and analysis of the foundation soil in the designated area and of methods for the design of

---

[1] International Conference on Collision and Grounding of Ships; July 2001 in Copenhagen; October 2004 in Tokyo
[2] The state of the art is documented by the presently valid German and international standards (DIN 1054; DIN 4014; DIN 4026; EAU 1999; American Petroleum Institute 1993; Germanischer Lloyd WindEnergy GmbH 1999; Det Norske Veritas Classification A/S 1992).

pile foundations. They also list the relevant and standards and recommendations valid at the time of publication.

All of the methods and procedures mentioned in Wiemann et al. (2002) are quasi-static approaches. The long-term behaviour of foundation soil, especially under cyclic and dynamic loads, has been the object of research for some time. However, it does not yet seem to be totally clear, which circumstances effect the softening or the hardening of different soil types, and whether this behaviour can be predicted in a numerically valid manner, i.e. without significant influence of the finite element mesh geometry on the results (Niemunis 2002; Maier 2002).

17.2.3 Collision of Ships and Offshore Wind Turbines

The numerical calculation of ship-OWT collisions is basically an extension of the calculation of ship-to-ship-collisions. While in ship bow-ship longside collisions, one of the elements (the bow of the colliding ship) can be assumed to be rigid (Kulzep and Peschmann 1999), in a ship-OWT collision, both structures are deformable. Additionally, the interaction of the construction and the foundation soil must be taken into consideration.

17.2.4 Numerical Modelling

The numerical model for the collision calculation consists of two main parts:

1. the Offshore Wind Turbine, consisting of the structure itself, the foundation, and the surrounding soil; and
2. the colliding ship and the surrounding water.

Four structures and media are modelled, each coupled with the other ones, i.e. forces and movements can be transmitted. The interfaces water/OWT and water/soil are neglected; forces are transmitted only indirectly from the ship to the soil via the OWT.

The two direct collision elements (ship and OWT) are idealised as finite element models, whereby the contact area is modelled in a more detailed manner than the remaining parts. For example, the nacelle of the OWT and the only indirectly affected parts of the ship are idealised as rigid bodies.

The consideration of the foundation soil as an elasto-plastic deformable body is necessary because, together with the foundation, it establishes a complex system not to be separated in a mechanical sense. Several loads are introduced to the soil, so that settling, vibrations or even loss of bearing capacity may occur (especially in such extreme load cases as collisions).

In order to obtain realistic results from the impact simulations, the geometry, mass, and stiffness needed to be accurately represented in the finite element model. To ensure realistic models to be constructed, structural plans for the ship types and the foundation designs were obtained from shipyards and wind park developers[3].

Foundations of Offshore Wind Energy Conversion Systems

As test models, a monopile, a jacket, and two tripod foundations (North Sea and Baltic Sea locations) were chosen. There are big differences as far as soil properties are concerned between the locations in the North Sea (sandy soil) and in the chosen Baltic Sea location "Sky 2000". Figure 1 shows the locations of the designated areas:

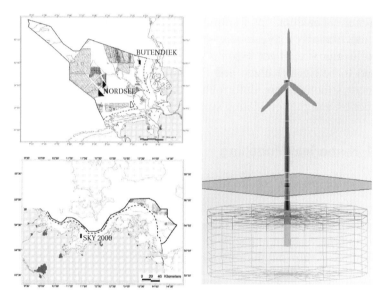

Fig. 1. Locations North and Baltic Sea   **Fig. 2.** Monopile

Monopile A monopile is the most cost-effective foundation type for offshore wind energy conversion systems. Due to its simple global design, it is the preferred solution in areas with water depths up to 25 m and soil consisting of mostly sand. Since it cannot bear great horizontal forces and moments because of its small lever arm, its global stiffness is generally

[3]   We would like to thank the IMS Engineering Company, F + Z Baugesellschaft mbH and Offshore-Bürger-Windpark Butendiek for their cooperation.

rather low, and does not develop much resistance in case of ship collisions. The OWT shown in Fig. 2 was designed for the "Butendiek" wind park in the North Sea, west of the island of Sylt, at a water depth of 20 m. The hub is located 80 m above sea level, and the foundation pile is 28 m long. The mass of the monopile is 300 t, the transition piece is about 124 t, and the overall mass is 665 t.

Jacket Jacket structures are widely used for offshore applications. A jacket combines high global stiffness with low structural mass. For offshore wind energy however, costs of manufacture and installation seem to be more relevant. It may be used at locations with greater water depths, e.g. more than 25 m and up to 50 m. Due to its large global and small local stiffness, it exhibits a large variation of failure modes during collisions. The jacket foundation used for the calculations was designed for "Butendiek", the same location as the monopile above. Although at that location there is a water depth of only about 20 m, so that jacket foundations might not be the most cost effective alternative, it was thought of a test site to evaluate this foundation type for offshore wind energy conversion systems.

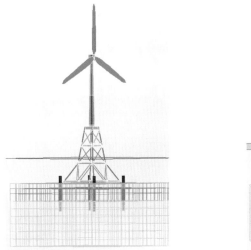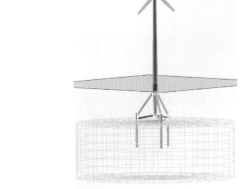

Fig. 3. Jacket and tripod structures

The hub height is 80 m above sea level, and the foundation piles are 22 m long. The mass of the jacket is 378 t, the four piles are 119 t, and the overall mass is 739 t which is 11 % more than that of the monopile structure.

Tripod Tripods are mostly used in areas with water depths greater than 25 m. The OWT shown in Fig. 3 was designed for a North Sea location with about 40 m water depth. The global stiffness of the tripod is comparable to that one of the jacket foundation, but consists of fewer components, which makes it easier to build, and decreases building costs. The local stiffness of the diagonals is much higher compared to those of the jacket, resulting in higher resistance to structural failure.

The hub height is 93.5 m above sea level and the tip of the foundation piles will be at 35 m below the sea bottom. The mass of the tripod is 473 t, and the three piles together have a mass of 331 t. The overall mass is 1,046 t. This is not comparable to the other structures, because it is designed for a different location with a much greater water depth.

Ship Types

In cooperation with Germanischer Lloyd, ship types were selected for the calculation of collisions. The decisive factor was the commonality of types. An analysis by Germanischer Lloyd on marine movement in the North and Baltic Seas was used as a basis (ISL 2000). The percentage of different ship types was ascertained for the years 1995 to 1999, and estimated for 2010 (Table 1). The absolute number of maritime movements rises from 373,023 during 1995-1999 (mean value) to 450,086 in 2010 (estimated).

Table 1. Percentage of individual ship types of total marine movements

Year	general cargo	ferries	tanker	container	bulker
1995 - 1999	5 %	24 %	12 %	8 %	5 %
2010	49 %	21 %	11 %	5 %	5 %

First, a 31,600 tdw double-hull tanker and a single-hull (150,000 tdw) tanker were selected. Later, a container ship (2,300 TEU) and a bulk carrier (170,000 tdw) were added. RoRo-ships were not taken into consideration (see Fig. 4 and Table 2).

DH: medium-sized double-hull tanker (31,600 tdw)

MH: large single-hull tanker (150,000 tdw)

CS: container vessel (2,300 TEU)

BK: bulker carrier (170,000 tdw)

Fig. 4. Different ship types used in the analysis

Table 2. Principal dimensions

Ship	double hull	single hull	container	bulker
Length [m]	150.00	304.00	250.00	283.00
Breadth [m]	27.72	46.00	29.80	46.00
Height [m]	17.30	23.30	16.50	24.40
Draught [m]	12.80	16.50	12.00	18.00
Deadweight [t]	31,600	150,000	--	170,000
Mass [t]	45,000	200,000	52,000	250,00

Calculation Procedure and Boundary Conditions

All calculations were done using LS-DYNA, a nonlinear explicit finite element programme widely used for crash applications. For static and quasi-static procedures, a nonlinear implicit solver implemented in

LS-DYNA was used. The movement of the ship was calculated by using a subroutine implemented in the code. It has already been introduced at the second ICCGS 2001 (Le Sourne et al. 2001).

The models were generated in MSC.Mentat, a pre-processor for the nonlinear finite element code MARC, and then the MARC input file (only nodes, elements, materials, boundary conditions and contact definitions) was translated into an LS-DYNA input deck which could easily be appended and modified by using a text editor to directly write LS-DYNA input.

Quasi-Static Pre-Calculation: Calculations were made using as many boundary conditions as necessary to obtain a realistic model. First, the different OWT models were loaded with static forces (gravitational and other):

Earth pressure at rest
Since there is no general estimate for earth pressures after driving a pile, earth pressure at rest was taken as a rough mean estimate.

Gravity force
Foundation structures and turbines were loaded with their assigned mass.

Working loads
Working loads were applied as quasi-static forces according to information given by the manufacturers.

Wind and wave forces were calculated according to Germanischer Lloyd (1999).

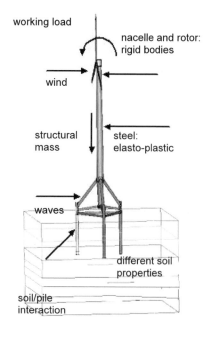

Fig. 5. Quasi-Static loads

After this static implicit calculation procedure, element stresses, nodal displacements and contact forces were written into a data file to make them available for the simulation.

Idealisation of the ships and their motion: The considered ship types are very long (up to 300 m) compared to the width of the OWT-foundation (up to 30 m). Therefore, the ships were not fully discretised. Only one or two holds were idealised as a finite element mesh. The rest of the ship was modelled as a rigid body connected to the mesh at the outer nodes of the longitudinal formation (see Fig. 6).

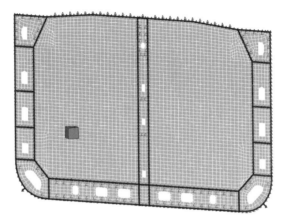

Fig. 6. Connection of deformable and rigid parts, connection nodes are shown as black lines

   In addition, the ship's motion before and during the collision must be taken into consideration. This is influenced by hydrodynamic forces. According to the theory of potential, the hydrodynamic forces can be easily determined in the harmonic agitation of the hull. The procedure has already been implemented in the present version of LS-DYNA (command *BOUNDARY MCOL). The calculating method used here is similar to the procedure in Petersen (1982). The components of movement {x(t)} are determined by the formula

$$\{[M] + [m(\infty)]\}\{\ddot{x}(t)\} + \int_0^t [g(r)]\{\dot{x}(t-\tau)\}\,dr + [K]\{x(t)\} = \{f(t)\}.$$

with

$$[g(\tau)] = 2\frac{2}{\pi}\int_0^\infty [C(\omega)]\cos\omega\,\tau d\omega$$

$[M]$ mass matrix
$[m(\infty)]$ Hydrodynamic mass matrix for the vibration frequency $\omega = \infty$
$[C(\omega)]$ damping matrix
$[K]$ Matrix of the hydrostatic restoring force
$\{f(t)\}$ External force vector

In this procedure, the coefficients of the differential equation are recalculated with every calculation increment that the subroutine is called.

Dynamic Simulation: For the calculations, the following boundary conditions must be set:

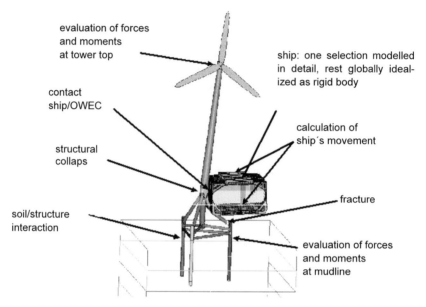

evaluation of forces
and moments
at tower top

ship: one selection modelled
in detail, rest globally idealized as rigid body

contact
ship/OWEC

calculation of
ship´s movement

structural
collaps

soil/structure
interaction

fracture

evaluation of forces
and moments
at mudline

Fig. 7. Collision model: input data, model features, and output data

Drift velocity for a ship with engine failure: Basically, a drift velocity of 2 m/s has been assumed, in individual cases, collisions with 3 or 4 m/s were simulated in order to be able test for differences in impact.

Variation of the moving direction: In standard cases, the initial velocity was set in the horizontal direction. For simple modelling of the sea's condition, a vertical component v_z of 0.5 m/s has been added in some cases.

Drift angle relative to the OWT: In case of a centric strike (the ship hits the OWT with its centre of gravity at a 90-degree-angle), the bulk of the kinetic energy is passed to the OWT, because little or no rigid body rotation, and hence energy consumption by damping by the surrounding water is actuated. This global worst case scenario does not necessarily produce the worst damage to the ship, because local effects may be the decisive factor.

Location of impact (OWT): Depending on the actual water level, different parts of the foundation structure may be struck by the ship. Analysis results will differ.

Expected Failure: Failure of individual modules is a decisive part of the simulation. There are a number of actuators for failure:

Plastic Strain (Ultimate Strength): In the constitutive model (piecewise linear plasticity) for steel, a failure criterion is implemented that considers an element as ruptured if a certain amount of effective plastic strain is reached. Kulzep and Peschmann (1998 and 1999) give values for certain element length/thickness ratios (see Fig. 8).

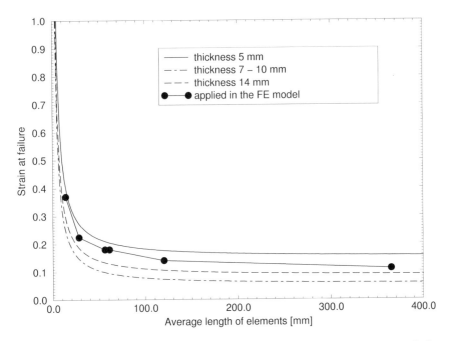

Fig. 8. Plastic strain at failure depending on characteristic element length and plate thickness

Buckling: Buckling is a type of collapse which can occur even within the elastic range. In compressive and bending loads, elements can be deformed in such a way that they will be greatly deformed, even at small extra loads. They become effectively instable. These phenomena are considered in the non-linear method. Nevertheless, these effects in dynamic processes should be evaluated qualitatively rather than quantitatively, because the mesh used is rather coarse.

Foundation soil: The immense load, which is passed to the soil, can lead to large-scale deformations. Heavy plastic deformations can only be brought about by modifications of the granular structure (e.g. shear bands), i.e. the soil loses its bearing capacity and collapses. Mechanisms of soil collapse cannot be represented in the present model because of its simplifications. Too much rigidity of the soil can lead to defective results in the calculation. This defect is however on the "safe side", because the OWT is able to offer more resistance, and tends to cause greater damage to the ship.

17.3   Results

In this section, mainly calculation results of collisions with the double hull tanker are presented, because it is easier to compare the effects of the ship impact on the different foundation types. Results of collisions with other ship types are presented briefly.

17.3.1 Monopile

The monopile considered (Fig. 9) cannot absorb the entire kinetic energy of the ship, i.e. the ship does not come to a full stop. With a mass of 45,000 t and an initial velocity of 2 m/s, the initial kinetic energy is 90 MJ. After the ship first comes into contact with the structure (1), buckling is induced in the pile by the impact. As the pile is pushed further away, it eventually fails at a transition between medium dense and dense sand (2). After a total displacement (ship) of about 5 m, contact is again lost (the monopile falls away from the ship).

At this point, 80 MJ of the initial kinetic energy of 90 MJ have been transferred. About 70 MJ have been absorbed by the OWT, of which 40 MJ have been passed to the soil by friction and pressure, and another 10 MJ effectively transformed into deformation energy within the ship (see also Fig. 9). At location 3 in Fig. 9, buckling develops due to the high mass of the turbine at the tower top. This may lead to another failure scheme if a 5 MW turbine of e.g. 450 or 500 t – instead of about 100 t, as here – is installed. Currently, OWTs with this high mass are being developed.

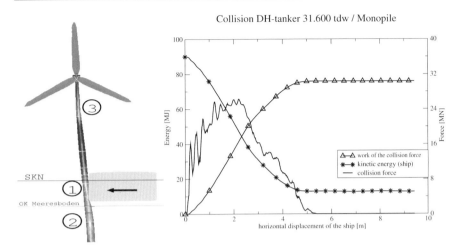

Fig. 9. Monopile: Failure mechanism and calculated collision force and energy

Considering the double hull tanker, only minor damage to the ship's structure has been observed. There was no scenario calculated with the inner hull penetrated. Although there has been only little plastic deformation, penetration of the outer hull is possible. According to Germanischer Lloyd's classification regulation for collision enhancements of ocean-going vessels, finite elements with plastic strain exceeding 5 % are to be considered broken. In this case the areas coloured black in Fig. 10 are considered to be destroyed, so that here, the outer hull has been penetrated.

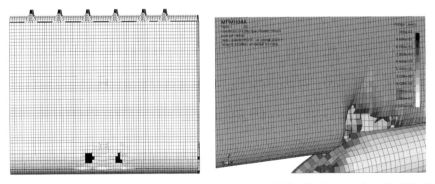

Fig. 10. Possibly penetrated parts of the outer hull (black) of the 31,600 tdw double hull tanker (left) and calculated penetration of the single hull tanker after 15 s (right)

Similar results were obtained with the 2,500 TEU container vessel. No serious threat to the environment could be identified for collisions with these two ship types.

There were no serious risks detected regarding collisions with the single hull tanker and the bulk carrier, except one scenario in which the tanker drifted onto the monopile at an angle of 60, with the centre of gravity not located at the geometric centre of the modelled part. Here, the ship slid along the pile, causing it to fail just at the contact area and tearing a hole into the side structure. This caused a great leak which – depending on the density of the fluid and the water pressure outside – might allow the cargo in two holds to completely spill into the sea and thus cause damage to the marine environment and high costs for removal (see Fig. 10).

Due to possible shortcomings of the finite element model, these results should not be seen as an exact representation of a real accident, but they may well show the possibility of occurrence and possible consequences of such accidents. These consequences should be the basis for preventive measures, which must be developed and implemented for each offshore wind farm (Kremser 2004).

17.3.2 Jacket

As already mentioned, the jacket structure is very stiff globally, but its member substructures are rather fragile. During a collision of the considered double hull tanker with the jacket, it destroys the substructure by first tearing it off its bearings and then penetrates the jacket about 10 to 14 m (see Fig. 11, (1)). As long as the ship is still moving, the jacket does not collapse, because the collision force "replaces" the bearing of one leg of the jacket. Once the entire kinetic energy of the ship is consumed, and it comes to a stop and the horizontal force on the jacket decreases, it may fail instantaneously depending on the distance the ship penetrates the jacket structure (also see Fig. 11 for energy calculations).

When the double hull tanker comes to a full stop, approx. 73 MJ will have been transferred by the collision force: 61 MJ as deformation energy into the jacket structure, and 12 MJ back into the ship. The difference of 17 MJ will have been consumed by the rigid body rotation of the ship. Because damping will have been included in the calculations, free rotation of the ship is not possible.

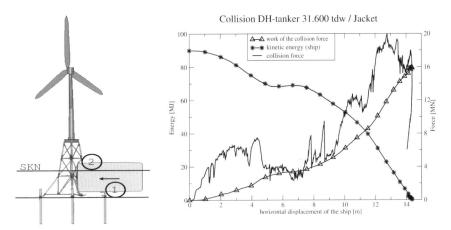

Fig. 11. Jacket: Failure mechanism and calculated collision force and energy

The calculated damage to the ship's hull is more severe compared to the monopile. Here too, only the outer hull of the double hull tanker is penetrated. In Fig. 12 the penetrated areas (deleted elements) and the elements exhibiting more than 5 % plastic strain are shown. The lower zone has been damaged by one leg of the jacket structure, whereas the upper hole results from contact to a joint. In general, there is no great hazard to the environment resulting from collisions of a ship of this size, when considering only the jacket foundation structure.

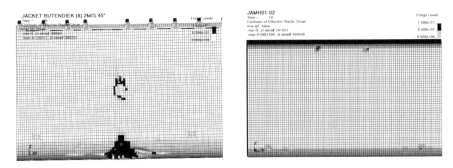

Fig. 12. Penetrated areas of the outer hull (black) of the 31,600 tdw double hull tanker (left) and collision with the 150,000 tdw single hull tanker: no penetration of the hull

Considering a ship of roughly four times the mass of the above one, the jacket foundation will simply be ripped away and collapse. In this case –

the single hull tanker drifts onto a structure perpendicular to one side of the structure – the jacket will fail at the connection between the structure and foundation piles. The turbine and the tower will fall in the direction of the ship, but the nacelle and the rotor would not impact the ship's deck, but rather fall into the sea behind the ship (Fig. 13).

Since the rotor and the nacelle have been modelled as a single entity with no disassemble criteria, no conclusions can yet be made regarding the risk of the actual turbine, or parts of it, falling onto the deck of a colliding vessel.

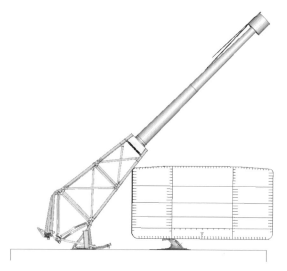

Fig. 13. Collapsing offshore wind turbine

17.3.3 Tripod

The tripod is first hit either at or above the central joint, or at one or two diagonals (1), depending on the draught of the ship and the location of the central joint. The ship then pushes the structure away, resulting in either one or two diagonals to tear off, consuming much of the ship's energy, or a pile to be pulled out of the soil for some distance (2), causing the structure to relax a bit, although it may also fail at a later point (Fig. 14). Smaller vessels, such as the double hull tanker may come to a full stop.

If the ship comes into contact with one diagonal, it will suffer severe damage at this point, because of the relatively high stiffness of the tube. Depending on sea conditions, not only the outer hull but also the inner hull of the 31,600 tdw double hull tanker may be penetrated (case 1).

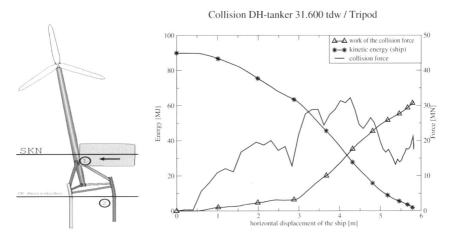

Fig. 14. Tripod: Failure mechanism and calculated collision force and energy

Provided the ship does not hit the diagonal strut, the collision event proceeds much like that with the monopile (case 2) and the consequences are similar to those of a collision with a monopile.

In case 1, 60 MJ were transferred from kinetic energy into deformation energy. The remaining 30 MJ was consumed by rigid body rotation of the ship (damping was included!). Only 16 MJ were consumed by the tripod, about 26 MJ by the soil, and another 18 MJ by the ship as deformation energy, which caused the damage.

In case 2, 80 MJ dissipated into the structures, but only 5 MJ were consumed by the ship, 42 MJ by the Offshore Wind Turbine, and 33 MJ by the soil. The difference between 5 and 18 MJ explains the difference in the damage showed in Fig. 15.

Comparing the internal (deformation) energy of the ship obtained from the calculation with the monopile with case 2, the tripod (5 MJ) performed even better than the monopile (approx. 10 MJ). The areas shaded black in Fig. 15 (plastic strain above 5 %) are about the same size as calculated in the monopile collision (Fig. 10). Placing the central joint deep enough to prevent it from having contact with the ship should be an acceptable measure to increase the passive safety of the foundation structure in order to satisfy regulations for the provision of safe structures.

Figure 15 shows a plot of an inner hull penetrated by the leg of the tripod structure. This event illustrates the general problem of providing safe structures, and proving safety by calculations. In any case there exists a (small) possibility of severe consequences in a collision event regardless which structure or which ship type is considered.

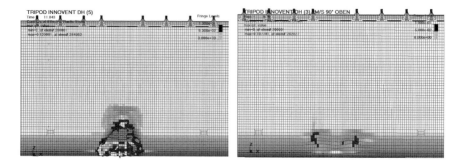

Fig. 15. Contact below central joint: Due to high stiffness of the diagonal strut even inner hull may be damaged (case 1). Contact above central joint: Almost no damage to the ship's hull (case 2)

In calculations with the other three ship types, similar behaviour could be observed. The calculations with the single hull tanker and the bulk carrier always led to leakage of one or more holds. Due to the low position of the penetration zone, large portions of the content of the affected holds may contaminate the marine environment.

A calculation with the container ship confirmed the aforementioned results. Here, damage was rather large. It is possible that during and after a collision, containers fall from the deck into the sea. As large holes in the ship's hull were observed, it might even be possible for containers to fall out of the holds. A calculation similar to case 2, above, with the double hull tanker was not conducted with the other ships.

17.3.4 Comparison

Comparing the energy absorption of the three foundation structures during the collision with the double hull tanker, it was shown that the monopile fails before the ship's kinetic energy of approx. 90 MJ is consumed. The monopile fails in any collision with one of the other ships. Due to its weakness, the damage to the side structures of these ships is relatively low.

Although the ship comes to a full stop in the collision with the jacket structure, damage is not much more severe than with the monopile. High global and small local stiffness allows the collision force to be transferred into the jacket over a long period of time, and through many contact areas. Collisions with a tripod foundation cause more severe damage to the ship:

If the ship drifts onto one of the diagonal legs, all the energy will be transferred through this contact point until the side structure of the ship reaches the OWT's tower. The relative strength of the diagonal and the tripod structure in general may lead to penetration of both the outer and inner hull of a tanker, as it was calculated here, and even more if a single hull vessel is considered. The probability of penetration and resulting leakage, with all its consequences, is probably higher than would be acceptable.

17.4   Recommendations

17.4.1 FSA: Risk Matrix

In this case, it is helpful to visualise the risk of severe environmental impact of ship-Offshore Wind Turbine collisions with a tool like the risk matrix proposed e.g. in (IMO 2002). We define four degrees of consequences and four grades of frequency of collisions. Here, we can give examples of choices of appropriate consequence grades for any of the examined foundation structures.

Consequences

In Table 3, the grades of consequences are defined in respect to environmental consequences because we only had to determine consequences for the marine environment in case of collisions. It should be thought of consequences to the OWT, the ship, and possible loss of lives as well.

Table 3. Definition of consequence grades

consequence grade	environmental damage
minor	no damage to the marine environment
significant	Operating supplies from wing tanks or tanks in the double bottom spill into the water; no structural damage to inner hull or double bottom
severe	One or more holds are penetrated: Cargo flows into the sea; inner hull or double bottom also penetrated
catastrophic	ship breaks apart and/or sinks

Probabilities

The probabilities (occurrence of collisions per year) are only a proposal according to Otto et al. (2002a). Probabilities have to be determined by the authorities; i.e. they have to decide whether a probability value is acceptable or not. The values below are merely an indication showing the differences between the three foundation structure types.

Table 4. Definition of frequency grades

frequency grade	probability [1/a]
frequent	$p > 2 \cdot 10^{-1}$
reasonably probable	$2 \cdot 10^{-1} \geq p > 2 \cdot 10^{-2}$
remote	$2 \cdot 10^{-2} \geq p > 2 \cdot 10^{-3}$
extremely remote	$2 \cdot 10^{-3} \geq p > 2 \cdot 10^{-4}$

Frequency grades have to be determined according to the actual location of a proposed wind farm. The combination of both grades yields a position in the risk matrix (Fig. 16). Here, the risk is quantified by numbers from 1 (very low) to 7 (very high). Risk numbers of 1 to 3 would be acceptable:

catastrophic	4	5	6	7
severe	3	4	5	6
significant	2	3	4	5
minor	1	2	3	4
consequences probability	extremely remote	remote	reasonably remote	frequent

Fig. 16. Risk matrix

According to the results of the calculations, monopile and jacket structures would be in consequence grade 2 (significant), which would allow collisions to occur remotely (less than once every 50 years).

The Tripod foundation would have to be placed in consequence grade 3 (severe), which would allow collisions to occur extremely remotely (less than once every 500 years) unless the condition is satisfied that no ship could come into contact with the central joint of the structure. Then, it could be placed in consequence grade 2 (significant).

17.4.2 Recommendations for Monopiles, Jackets and Tripods

Monopile foundations exhibit the lowest risk in case of collision. If the probability of occurrence of a ship-monopile collision is not too high – which should be achievable – no measures have to be taken into consideration to enhance passive safety for a monopile.

The work of the collision force can be transformed into large deformations within the **jacket structure** as far as the structure is able to withstand the ship long enough without being torn off.

Local damage in the model caused by the jacket's joints should not lead to widely damaged areas of the ship's hull. As the joints have been modelled very stiff, this condition leads to greater damage to the ship than would probably occur in a real collision.

Here too, no measures to enhance passive safety have to be taken, provided the probability is low.

If a ship hits the diagonal chord and the central joint of the **tripod**, severe consequences may occur. To minimise these risks, two points should be observed if tripods are planned:

1. The central joint of a tripod should be located lower than the maximum draught of any ship travelling regularly in the area.
2. If the water depth is not more than 25 m, a monopile should be taken into consideration. If the sea is deeper, a jacket could be the appropriate solution.

17.4.3 Measures to Increase Active Safety

The risk of collision can be reduced, but it cannot be totally avoided. In addition to the "safe" structures and the evaluation of collision risks, the risk management for each wind farm should include two goals:

1. minimising the collision risk by observing and controlling ship traffic: radar, optimising ships for collision safety, and training of the crews of ships;

2. Countermeasures: scenarios that might lead to collisions should be compiled, and strategies developed to avoid them (e.g. by the use of tug-boats that have to be always available at the site).

17.5 Conclusions

The purpose of this paper is to investigate the behaviour of three foundation structures of offshore wind turbines (OWTs) in collision scenarios. Three different foundation types, monopile, tripod, and jacket, were subjected to finite element analysis. LS-DYNA was used to predict the damage caused by collisions of four different ship types: single and double hull tankers, bulk carriers, and container ships. The main focus was placed on the energy absorption of the different structures, and on the possible effects on the marine environment of the predicted damage.

The very critical scenarios are collisions of any ship type with a tripod foundation structure whose legs are in contact with the ship. This would cause major local deformations of the ship's hull, resulting in penetration of the outer hull, and, depending on sea conditions, even of the inner hull.

Other scenarios yielded less critical results. It was demonstrated that monopile and jacket structures are safer than certain tripod structures. One way to measure safety was to compare the collision energy absorbed by the ship. It was always much lower than the energy absorbed by the foundation structure and the soil, except in the very critical scenarios mentioned above.

One way to solve the problems with the tripod foundation is to change the design in order to place the central joint at a deeper point, so that a ship will hit neither the joint nor the leg. Calculations have shown that this kind of construction behaves more like a monopile. Damage to the ship was in the same order of magnitude.

It is not always possible to change the design of the tripod as described above, e.g. if the sea is not deep enough for a design that ensures stability. Another possibility is to change the design of the foundation completely.

The monopile and the jacket designs are best suited to the depth ranges of 0 to 25 m and 20 to 50 m, respectively. Although foundations of Offshore Wind Turbines can be chosen to ensure passive safety, there is another risk which has not been examines. Even in case where the foundation does not fail completely, the actual turbine of the OWT may not stay attached to the tower. Due to the high mass which varies between approx. 100 t for a 3 MW turbine and 450 t (!) for a 5 MW turbine, the forces and moments at the tower top will be very great, and the bearing may fail.

Whether the nacelle, or parts of it, miss the ship or fall on its upper deck is difficult to determine and depends on the scenario.

References

American Petroleum Institute (1993) Recommended Practice for Planning, Designing and Constructing Fixed Offshore Platforms – Working Stress Design (API RP 2A – WSD), 20th Edition. Dallas

Braasch W, Nusser S, Jahnke T (2001) Risikoanalyse Offshore-Windenergiepark Borkum West (Version 1). Germanischer Lloyd Offshore and Industrial Services, Hamburg

Clauss G, Lehmann E, Östergaard C (1998) Meerestechnische Konstruktionen. Springer, Berlin Heidelberg New York

Det Norske Veritas Classification A/S (1992) Foundations, Classification Notes No 30.4. Høvik

DIN 1054 (Entwurf 2000) Baugrund; Sicherheitsnachweise im Erd- und Grundbau. Beuth

DIN 4014 (1990) Bohrpfähle; Herstellung, Bemessung und Tragverhalten. Beuth

DIN 4026 (1975) Rammpfähle; Herstellung, Bemessung und zulässige Belastung. Beuth

EAU (1990) Empfehlungen des Arbeitsausschusses "Ufereinfassungen" Häfen und Wasserstraßen, edn 9. Ernst und Sohn, Berlin

Egge ED, Böckenhauer M (1989) Berechnung des Kollisionswiderstandes von Schiffen sowie seine Bewertung bei der Klassifikation. In: Jahrbuch der Schiffbautechnischen Gesellschaft, 83. Band. Springer, Berlin Heidelberg New York

Ferry M (2001) MCOL User's Manual. Principia Marine, Nantes

Ferry M (2001) MCOL Theoretical Manual. Principia Marine, Nantes

Germanischer Lloyd (2002) Klassifikations- und Bauvorschriften I Schiffstechnik, Teil 1 Seeschiffe.

Germanischer Lloyd WindEnergie GmbH (1999) Rules & Regulations, IV Non-Marine Technology 1999. Hamburg

ICCGS (2nd International Conference on Collision and Grounding of Ships) (2001) Proceedings. Copenhagen

IMO (International Maritime Organisation) (1997) Formal Safety Assessment: Interim Guidelines for the Application of FSA to the IMO Rule-Making Process, IMO MEPC 40/16. London

IMO (International Maritime Organisation) (2002) Guidelines for Formal Safety Assessment (FSA) for Use in the IMO Rule-Making Process, IMO MEPC 392. London

ISL (Institut für Seeverkehrswirtschaft und Logistik) (2000) Statistische Daten zu Schiffsverkehren in Nord- und Ostsee. Bremen

Kremser (2004) Risk assessment and precautionary measures for offshore wind parks. In: Proceedings of Scientific Forum of the Federal Ministry for the Environment, Nature Conservation and Nuclear Safety (BMU) on Offshore Wind Energy Utilisation, Berlin

Kulzep A, Peschmann J (1998) Grundberührung von Doppelhüllentankern, Abschlussbericht zum BMBF Verbundforschungsvorhaben Life Cycle Design, Teil D2. TU-Hamburg-Harburg, Hamburg

Kulzep A, Peschmann J (1999) Seitenkollision von Doppelhüllenschiffen, Abschlußbericht zum BMBF Verbundforschungsvorhaben Life Cycle Design, Teil D2A. TU-Hamburg-Harburg, Hamburg

Le Sourne H et al (2001) External Dynamics of Ship-Submarine Collisions. In: Preprints of 2nd International Conference on Collision and Grounding of Ships, Copenhagen

Livermore Software Technology Corporation (1998) LS-DYNA Theory Manual. Livermore

Livermore Software Technology Corporation (2003) LS-DYNA Keyword User's Manual Version 970. Livermore

Maier T (2002) Numerische Modellierung der Entfestigung im Rahmen der Hypoplastizität. PhD Thesis, Universität Dortmund

Niemunis A (2002) *Extended hypoplastic models for soils.* Post-doc. dissertation, Ruhr University of Bochum

Otto S, Nusser S, Braasch W (2002) Kollisionsrisiko von Schiffen mit Windenergieanlagen und die Gefahr der Belastung der Küstenregion. Germanischer Lloyd Offshore and Industrial Services, Hamburg

Otto S, Nusser S, Braasch W (2002a) Methoden zur Berechnung Kollisionsrisiken von Schiffen mit Windenergieanlagen. Germanischer Lloyd Offshore and Industrial Services, Hamburg

Petersen, MJ (1982) Dynamics of Ship Collisions. Ocean Engng 9:295–329

Richwien W, Lesny K, Wiemann J (2002) Tragstruktur – Gründung. In: Bau- und umwelttechnische Aspekte von Offshore Windenergieanlagen, pp 56–73. GIGAWIND, Hannover

Wiemann J, Lesny K, Richwien W (2002) Gründung von Offshore-Windenergieanlagen – Gründungskonzepte und geotechnische Grundlagen. Universität Essen, Essen

Yu X (1997) Strukturverhalten mit großer Verformung bis zum Brucheintritt und mit dynamischer Zusammenfaltung. Dissertation, Bericht No 579, Institut für Schiffbau der Universität Hamburg

Zhang S (1999) The Mechanics of Ship Collisions. PhD Thesis, Technical University of Denmark, Department of Naval Architecture and Offshore Engineering, Lyngby

Planning Aspects

18 Environmental Impact Assessment in the Approval of Offshore Wind Farms in the German Exclusive Economic Zone

Julia Köller, Johann Köppel, Wolfgang Peters

18.1 Introduction

Offshore wind farms are a renewable, but not entirely conflict-free form of power generation. Not only the effects on competing maritime uses, but also the marine environment must be taken into account in the authorisation of such wind farms. Particularly the interests of environmental protection and conservation have special weight, due to international and national legal provisions and agreements. For the approval of offshore wind farms in the German Exclusive Economic Zone (EEZ), the EU's Environmental Impact Assessment (EIA) directives and, if NATURA 2000 areas are involved, the appropriate assessment according to the Habitat Directive, are legally stipulated. In future, in the context of the spatial planning control of wind energy use, the Strategic Environmental Assessment (SEA) will also be mandatory in the German EEZ. These instruments have the task of ascertaining the impact on the marine environment caused by the construction, installation and operation of offshore wind farms, and of integrating this information into the decision-making process on the authorisation of such projects.

For this purpose, the procedural specifications and methodological requirements of the instruments for Environmental Impact Assessment and management must be adapted to the specific situation of the authorisation of offshore wind farms in the EEZ. This involves on the one hand the legal specifications of the authorisation procedure according to Marine Facilities Ordinance (SeeAnlV). On the other, this must be adapted to the specific conditions of the marine and to the specific correlations of impacts. Only then will an effective authorisation procedure be possible, in which the interests of the marine environment can properly be taken into account.

The research project being implemented at the Berlin University of Technology (TU Berlin) for the integration of environmental assessment instruments into the authorisation procedure for offshore wind farms has had the goal of establishing these required adaptations[1]. For this purpose, indicators have been developed to serve as a guideline, and to structure both the process of the EIA and the process of the appropriate assessment according to the Habitats Directive with regard to the specific legal specifications and assessment standards.

As a first step, the legal assessment standards were derived and developed. This step involved both the determination of the protected assets relevant under the authorisation process, and also the definition of the relationship of effects between the project and the protected assets concerned, which would be relevant to the decision-making process and were to be considered in the context of impact prognosis. For every correlation of impacts, the scientific state of knowledge is explained, and initial attempts for the assessment of possible effects of offshore wind energy turbines introduced and discussed. The core item of the project results is a discussion platform, in which the central EIA and authorisation-related issue of evaluation of environmental effects is described for various correlations of impacts (Köppel et al. 2003).

18.2   Legal Standards for the Assessment of Environmental Impacts in the Approval of Offshore Wind Farms

Under § 2 of the Marine Facilities Ordinance (SeeAnlV), wind farms in the EEZ require the approval of the Federal Maritime and Hydrographic Agency (BSH). Such approval is a "bound decision"; that means that approval can only be denied to the applicant if reasons for refusal under § 3 SeeAnlV are present.

Accordingly, approval must be refused "if [...] the marine environment is endangered without there being any possibility of such endangerment being prevented or compensated by time limitation, conditions or stipulations. Grounds for rejection shall in particular be present if [...]

- pollution of the marine environment as per Article 1, Sect. 1 No. 4 the United Nations Convention on the law of the Sea of 10 December 1982 is to be feared; or if
- bird migration is jeopardized"(cf. also Ch. 1.3).

---

[1]   The research project was funded by the Federal Ministry for the Environment, Nature Conservation and Nuclear Safety.

The EIA is a non-autonomous element of the admission process. It formulates no standards of its own for the assessment of environmental impacts as a basis for decision-making on the authorisation of projects. Hence it must orient itself towards the grounds for refusal formulated in the Marine Facilities Ordinance. The Environmental Impact Assessment hence has the task of assessing the effects of offshore wind farms for the marine environment, and evaluating them in terms of endangerment.

18.3   Demands upon the Environmental Impact Assessment in the Context of the Authorisation Procedure

In the context of EIA, the direct and indirect effects of a project are to be identified, described and assessed. In this context, the focus is to be on facts relevant to the decision-making process. Therefore, in the approval of offshore wind farms, the question as to which aspects the Environmental Impact Assessment is to particularly address will not be determined solely by the fact that there may be impact correlations between the project and an environmental factor. Rather, the question to be addressed is whether the impairment of the marine environment resulting from the possible correlation of impacts could be so considerable that a situation of "endangerment of the marine environment" as defined under the Marine Facilities Ordinance would be present. Not every project-induced change, deterioration or impairment would therefore constitute an endangerment to the marine environment and to bird migration, such as would necessarily lead to rejection. The Environmental Impact Assessment must focus its contribution to the decision-making process in accordance with the standards of the authorisation procedure. Accordingly, only those correlations of impacts are to be included in the assessment which could demonstrably cause such considerable adverse environmental consequences as to constitute a situation of endangerment of the marine environment, and would thus in fact be relevant for the approval decision.

For the identification of correlations of impacts which could possibly be relevant for the decision-making process, it is first of all necessary to ascertain which impact factors affect which protected assets, and which negative changes might arise as a result, which could ultimately cause a denial of approval. The concrete question to be asked is: which factors of the marine environment are affected by the typical impacts of offshore wind farms, and which factors of the marine environment must hence be considered as protected assets in the environmental assessment?

The protected assets which must in principle be taken into account in the ascertainment, description and assessment of environmental impacts are described in § 2 Sect. 1 of the Environmental Impact Assessment Act (UVPG). Since the approval for a wind farm is granted in accordance with the assessment standards stated in the Marine Facilities Ordinance (SeeAnlV), these protected assets are to be juxtaposed to the terms "marine environment" and "bird migration", and concretised in according with their specific expression in the EEZ. For example, for the protected asset "fauna" as defined under the Environmental Impact Assessment Act, the groups of species of animal living in or on the sea must be specifically named. Thus, for an Environmental Impact Assessment in the context of an approval process for offshore Wind farms, these following protected assets can be identified for the marine ecosystems of the North and Baltic Seas (cf. Table 1).

Table 1. Protected assets under the EIA Act, and definition of protected assets as per SeeAnlV

Protected Assets under the EIA Act	Specification of the "marine environment" and the "bird migration" protected assets, as stated in SeeAnlV
Human beings	Human beings
Fauna	Sea birds, migratory birds, marine mammals, fish, benthos
Flora	Esp. growth on structures, macro-phytobenthos
Soil	Seabed, sediment structure
Water	Seawater, hydrology
Air	Air
Climate	Climate
Landscape	Landscape
Interactions	Interactions between these factors
Material assets and cultural heritage	Wrecks

18.3.1 Effects of Offshore Wind Farms on the Marine Environment

In addition to the interpretation and definition of protected assets to be considered in relation to the marine environment and the Marine Facilities Ordinance, an understanding of the construction, installation and operation-related effects of wind farms is decisive for the effect prognosis to be developed in the context of the Environmental Impact Study. Impact factors, such as acoustic emissions, vibrations or barrier effects will arise from the specific qualities of the project elements, and can be deduced from the conceptual and technical design of the offshore wind farm. It is important to distinguish between impact factors and effects. An impact factor is a result of a project, and does not, in and of itself, constitute an impairment. An effect, by contrast, is characterised as a positive or negative change of one or more of the protected assets, caused by the impact factors. For the effect prognosis, those impact factors of an offshore wind farm are significant which cause negative effects on the marine environment or bird migration, and therefore could lead to denial of approval. Table 2 below shows an example of the installation-related impact factors of an offshore wind farm.

Table 2. Examples of installation-related impact factors of offshore wind farms

Installation-related impact factors include all those connected with the installation and its structures

Cause/place of effect	Possible impact factors	Biotic and abiotic protected assets affected
By foundations, masts, rotors and possibly transformer stations	Demands on area and space (soil, water, air); habitat loss	Birds, marine mammals, fish, benthos, seabed, landscape
By foundations and piles	Creation of artificial hard substrate, sealing	Fish, benthos, seabed, cultural assets (wrecks, archive function of soil)
By masts and rotors	Obstacle and barrier effect, fragmentation (collision danger); shading	Birds, fish, benthos
By facility lighting for identification (safety for navigation)	Artificial illumination	Birds
Small-scale, in the area of single wind turbines, and large-scale, due to an entire wind farm (due to masts)	Flow change	Hydrology (thermal water stratification, salinity, temperature, density, nutrient, pollutants)
At foundations (lack of scour protection)	Scouring	Seabed
Due to flow changes	Sediment swap	Seabed; fish close to floor
Safety-related fishing ban in the wind farm and the cable route	Reduction of fishing	Fish

18.3.2 Derivation of Effect Correlations Relevant for the Decision-Making Process

Whether an impact factor will cause an impairment (negative consequence) for one or more protected assets depends on the one hand, and decisively, on its intensity, and scope, and on the other on the specific sensitivity of the particular protected asset to this impact factor. By correlating the concretised protected assets of the marine environment and the impact factors emanating from the wind farm in a matrix, it is possible to identify the elements and protected assets of the marine environment which are not impacted by offshore wind farm operation. Certain elements of the marine environment are either so little sensitive to the impact factors, or they play such an insignificant role in the EEZ's of the North and Baltic Seas, that considerable impairment of these elements of the marine environment or of bird migration can be precluded from the start, according to the present state of knowledge. This involves particularly the protected assets climate, air and, material assets and cultural heritage. Given the present state of knowledge, endangerment as per the Marine Facilities Ordinance can also be precluded for the protected assets human beings, flora, and soil. These assets are shaded grey in Table 1. In the context of an EIA focused on the issues significant for the decision-making process, the detailed recording and assessment of these protected assets and elements can be dispensed with.

The correlation/combination of the protected assets and their specific sensitivity towards the impact factors caused by offshore wind farms not only shows which protected assets can be neglected in the context of the decision-making process. Also those correlations of impacts become identifiable which could fulfil the grounds for refusal and should therefore be examined thoroughly in the course of the Environmental Impact Study. Such correlations of impacts relevant for the decision-making process include:

- Displacement, disturbance or collision of sea birds due to construction activities and operation of the turbines;
- collision or diversion of migrating birds due to construction activities and operation of the turbines;
- damage to or displacement of marine mammals due to construction and operational noise;
- damage to fish fauna by sediment dispersion, vibration, or electromagnetic fields;

- change of character of fish occurrence due to introduction of new habitats (artificial hard substrates);
- damage to benthos communities by over-building, sediment exchange and changed benthos communities due to introduction of hard substrates;
- disturbance of the stratification of water, especially in the Baltic Sea;
- visual impairment of the landscape;
- maritime pollution due to ship collision.

Authorisations already issued confirm that the effects on some protected assets are negligible. However, there are other correlations of impacts towards which special attention is directed[2]. The correlations of impacts identified as relevant to the decision-making process are specified and differentiated (cf. Table 3).

During the course of the research project, the impact correlations and partial correlations listed in the Table 3 were checked and evaluated for their relevance to the decision-making process via detailed reviews of the literature on the current level of knowledge and information, and via discussions with experts in the respective areas. As a result of this detailed analysis, some of these correlations or partial correlations of impacts must now be classified as not relevant to the decision-making process. E.g., some expected effects appear at closer consideration as temporary and/or of such minimal extent that the impairment must be assessed as generally insignificant. In other cases, knowledge and information of the effects are still so limited that no conclusive statements on the possible extent of impairment which might permit an assessment of endangerment to the marine environment can be made. These correlations of impacts, too, are at present irrelevant for the decision-making process.

In sum, only six correlations of impacts can ultimately be considered as relevant to the decision-making process in the context of the authorisation procedure under Marine Facilities Ordinance, and hence to be pursued with special care in the context of the Environmental Impact Assessment (cf. shaded fields in Table 3).

---

[2] Cf. alternatively authorisation certificate of 23 Aug. 2004 for the Sandbank24 offshore wind farm, and rejection certificate of 20 Dec. 2004 for the Pommeranian Bight offshore wind farm, by the BSH.

Table 3. Applicable correlation of impacts and partial correlation of impacts relevant to the decision-making process

Protected asset	Correlation of impacts relevant to the decision-making process	Partial correlation of impacts relevant to the decision-making process
Seabirds	Displacement, disturbance or collision of sea birds by construction activities and operation of turbines	Temporary habitat loss due to the displacement effect of construction and maintenance traffic
		Permanent loss of habitat by displacement and barrier effects of wind farms
		Direct loss of sea birds by collision (bird strike)
Bird migration	Collision or diversion of migrating birds by construction activities and operation of turbines	Endangerment the bird migration by bird strike
		Endangerment the bird migration by the barrier effect of the wind farm
Marine mammals	Damage to and/or displacement of marine mammals by construction and operational noise	Lethal damage to marine mammals by noise emissions from ramming
		Displacement of marine mammals due to operational noise of turbines
		Displacement of marine mammals by ship and air traffic for maintenance
		Permanent damage to and/or displacement of mother-calf groups by noise emissions due to ramming
		Damage to and/or displacement of mother-calf groups by turbine noise
		Damage to and/or displacement of mother-calf groups by ship and air traffic for maintenance
Fish	Damage to fish fauna by sediment dispersion, vibration, electromagnetic fields and changes in fish occurrence by introduction of new habitats (artificial hard substrates)	Change of fish occurrences by introduction of new habitats artificial hard substrates)
		Damage to fish fauna by sediment dispersion during ramming or due to washing in of submarine cables
		Damage to or displacement of fish by vibration during the operation of the turbines
		Barrier effect due to electromagnetic fields of the feeder cables to the mainland

Table 3. (cont.)

Protected asset	Correlation of impacts relevant to the decision-making process	Partial correlation of impacts relevant to the decision-making process
Benthos	Damage and/or loss of benthic communities through over-building, sediment exchange by introduction of hard substrates	Space needs of turbines, possible elimination of benthic communities or species
		Changes of the species composition due to introduction of artificial hard substrates, and/or change of sedimentation (flow)
		Construction-related damage to benthic communities by sediment dispersion
Seawater, Hydrology	Marine pollution due to ship collisions	--
	Disturbance of the stratification of the water esp. in the Baltic Sea	--
Landscape	Visual impairments of the landscape	Temporary disturbance of the landscape by flight and shipping during construction activities
		Inherent long-term visual impairment of the landscape from the coast

Permanent habitat loss for sea birds

Wind farms can cause widespread avoidance of these areas by birds, resulting in a displacement effect. Particularly such for food-searching, disturbance-sensitive seabird species as red-throated diver and black-throated diver, this displacement, or the fragmentation of coherent ecological units (e.g. resting and feeding areas), can cause permanent habitat loss. Given a lack of alternative habitats, this can directly affect population development.

Endangerment of bird migration due to bird strike

Immediate endangerment to bird migration exists due to the risk of collision with the turbines (bird strike). Particularly during unfavourable weather situations, species which migrate at the height of the effect of the turbines (0 - 150 m) can suffer considerable collision rates. This loss of individuals can have a negative effect on the overall population development at some species.

Endangerment of bird migration due to the barrier effect

A wind farm in the German Exclusive Economic Zone (EEZ) can have a barrier effect on migrating birds, forcing them into energy-consuming evasive action. These unintentional detours can lead either to direct loss of individuals, or reduce reproduction rates, and hence have a negative impact on the population development of some species.

Hearing damage to and/or displacement of marine mammals

With regard to marine mammals, the effects of noise are to be expected mainly during the construction phase. Especially during the ramming of the foundations, considerable acoustic emissions can be of great significance for marine mammals. The noise immission during the construction period is not long-lasting, compared with the operational noise of the facilities, but for the species of mammals relevant in the North and Baltic Sea areas, it could lead to displacement or even to permanent damage (change in hearing thresholds, fatal injuries), depending on noise intensity and frequency range. Mother-calf groups are considered especially sensitive.

Marine pollution due to ship collisions

Offshore wind farms are an obstacle to shipping, and therefore constitute a potential risk of accidents. Collisions of ships with wind farms could occur either with or without pollutant spills (oil, other pollutants), resulting in considerable effects on the entire ecosystem, including the coasts. There is a high risk potential of collision in the areas of traffic separation. Since the intensity of impairment of the marine environment due to ship collisions is independent of its location and the specific nature of environmental assets occurring there, or their sensitivity, only the collision risk itself is relevant to the decision-making process in the context of the EIA.

Impairment of the landscape

Even if the marine area of the EEZ does not border directly on the coast, it constitutes an essential component of the coastal landscape, due to its long-distance effects. Major areas of the North and Baltic Sea coasts and the offshore islands are used by tourists to a considerable degree, and a clear view of the sea is an important factor for all resorts along the coast. The construction of wind farms introduces vertical structures into a space which is as a rule free of obstacles and characterised almost exclusively by horizontal structures. The installation of offshore wind farms can disrupt the tourist attractiveness of vacation sites and thus degrade the recreation experience to some degree. This visual impairment is particularly significant in regions where tourism is important.

The possibility that further research could uncover additional impact correlations which could lead to considerable impairment, and thus gain relevance to the decision-making process cannot be ruled out. However, in the interval it is possible to concentrate on the impact correlations listed, in the context of the Environmental Impact Assessment (EIA).

18.3.3 Prognosis and Assessment of the Effects of Offshore Wind Farms

An evaluation of the concrete impairment of marine ecosystems due to the correlations of impacts of planned offshore wind farms identified as relevant to the decision-making process requires knowledge on the one hand of the specific impact factors, and on the other, of the existing characteristics of the protected assets of the marine environment, both living and non-living, present in the project area. A recording and assessment of these environmental assets is a prerequisite for a significant prognosis of effects, and should be carried out in a target-oriented and impact factor related manner in terms of the effect analysis.

The standard programme for environmental examination of the BSH, which is binding and has been developed as a methodological foundation for the basic examination for project-accompanying monitoring, stipulates thematic and technical minimum requirements for the recording of the existing environmental situation. The monitoring programme has the goal of observing the effects of the construction, operation and dismantling phases on as broad a basis as possible (BSH 2003). The baseline survey thus stipulated and the results thus obtained at the same time represent the basis for the assessment of the technical significance of the protected assets. However, the standard programme for environmental examination, with its examination programme tailored towards monitoring, goes far beyond the drafting of a decision oriented Environmental Impact Study (EIS), in which ultimately only those aspects are to be shown which serve to elucidate the effect structures relevant to the decision-making process.

18.3.4 Prognosis of the Effects

While in the course of the baseline surveys examinations, good knowledge of the local characteristics of the protected assets is gained, knowledge of the possible effects of the construction and operation of offshore wind farms will still be incomplete. Experience gained in the context of other offshore projects, particularly oil and gas platforms, is helpful. Also, considerable knowledge gained ashore on the impact of wind farms can provide important information (e.g. species-specific avoidance behaviour towards the rotors, by birds). However, no direct transfer of knowledge is possible, due to the specific conditions in and above the sea.

The incomplete knowledge and information situation is a known problem for effect prognoses in the context of Environmental Impact Studies (EIS), because the EIA Act demands identification, description and assessment of expected environmental effects. A forecast is particularly difficult if there is no transferable knowledge about correlations of impacts. However, since an applicant has a right to a decision on authorisation, statements as to the expected effects are necessary, even if knowledge on the correlations of impacts is still incomplete. The pressure of this necessity is all the greater due to the construct of the "bound decision" (cf. Chap. 18.2). In principle, the developer has a right to authorisation of the project if it meets the legal requirements. The result is that the BSH, as the responsible authorisation authority, must approve offshore wind farms in the German EEZ, as long as there is no information indicating substantial endangerment to the marine environment.

In sum, the knowledge situation may fall into any of the following three classes:

1. It is known that there are correlations of impacts. There is scientific information indicating that the effects could be so serious that consequences for the decision-making process may result (endangerment of the marine environment).
2. It is known that there are correlations of impacts. So far however, scientific investigations have yielded no indications that these could be so considerable and endangering as to have consequences for the decision-making process.
3. It is conceivable that correlations of impacts could result. To date however, there is no scientific knowledge and methodology available to determine how to ascertain them.

Cause-Effect Chains

It is important to structure and process available information in such a way that a prognosis of possible effects with regard to the significance of impairments and possible endangerment of the marine environment is ascertainable in spite of knowledge gaps.

For a prognosis of the environmental effects, the single steps: effect analysis, effect prognosis, assessment, and decision-making, must be clearly separated. As a basis for effect prognosis, it is first of all necessary to analyse the structure of effects between the project and the various protected assets and factors of the marine environment (effect analysis) (Köppel et al. 2004). The cause-effect chain, which structures the correlations of impacts, can serve as a basic model for the analysis of effects (cf. Fig. 1). It is based on the principle that impact factors will be caused by a project, which can have an effect on the single components of the ecosystem, and may cause changes (effects) there.

Fig. 1. Basic model of a cause-effect chain (cf. e.g. Bernotat 2003)

When structured according to this model of the cause-effect chain, the researched and relevant information for every correlation of impacts relevant to the decision-making process is documented in the above discussion platform. For the authorisation decision, it is necessary to refer the identified and forecast effects of a project to the legally defined reasons for rejection. For this purpose, thresholds are defined which, if they are exceeded, can cause the impairments to be classified as considerable, and thus meet the criteria for rejection (cf. also Chap. 18.2). Then, based on this basic model of the cause-effect chain, the possible evaluation criteria can also be derived, and adequate evaluation procedures and methods developed.

18.3.5 Assessment of the Effects

The assessment of the forecast environmental impacts is a central component of Environmental Impact Assessment (§ 12 EIA Act). An initial assessment of the environmental effects by experts working on behalf of the developer will frequently already have been carried out in the context of the Environmental Impact Study. If this has been done in a manner appropriate to legal assessment standards, the authorisation agency can use it as a basis for evaluating the significance of the impairment and thus judge whether the criteria for "endangerment of the marine environment" have been met and the project may hence be inadmissible. In practice, the expert assessments in the EIS's have to date been inadequately oriented toward the legal assessment standards, so that the authorisation agency hardly considers them at all in its decision-making process (cf. Morkel 2005).

Complex assessment procedures should formally and in substance be structured and comprehensible. This thus requires that results which are as objective, differentiated and comprehensible as possible be provided. In the context of the authorisation procedure of offshore wind farms and drafting of Environmental Impact Assessments, several assessment steps must be undertaken:

If a change in the marine environment is classified as an impairment, this already presupposes a first assessment step (cf. Fig. 2). This means that a desired ideal condition must be defined in the form of environmental targets, which will serve as a standard for the degree of deviation of the impairment caused by the project. Theoretically, changes which, measured according to environmental targets, would be evaluated as positive, are also conceivable.

For the assessment of impairment intensity and hence for the assessment of significance, statements of target, against which the forecast changes can be assessed as deviations, must be present. For the maritime area, such environmental quality targets may be formulated both at the national level (e.g. in the context of protected-area designations) or in the context of international agreements (e.g. OSPAR-Convention). It is important that the assessment of the impairment intensity on the one hand, and the assessment of significance on the other, be understood as two different assessment steps. The assessment of impairment intensity measures the deviation from the target condition. The assessment of significance on the other hand refers to the socially tolerated degree of deterioration. The first assessment step, which can be carried out by the scientists, is a technical one.

The second step however is a legal one, carried out in the context of the environmental assessment according to § 12, EIA Act, and in which the authorisation decision the existing legal situation, in this case the rejection of approval, must take into account (cf. Fig. 2).

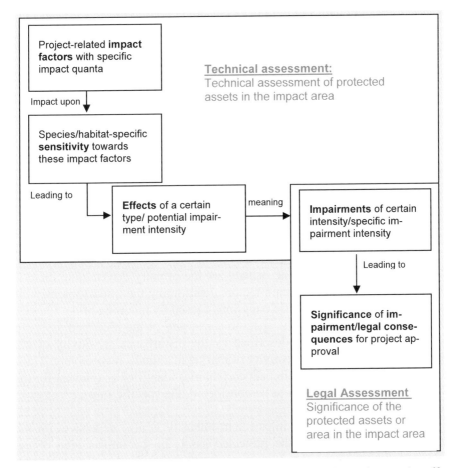

Fig. 2. Basic concept of the assessment of the significance of impairments by offshore wind farms in the context of the EIA

The intensity of the impairments to be assessed is determined by various factors and impact quanta. Besides the intensity and scope of the impact factors caused by the offshore wind farm, the sensitivity of the protected assets concerned, and their concrete spatial character, also affect the significance of the impairment. Finally the technical and legal significance of the protected assets or the area must be included in the scope of impact (cf. Fig. 3).

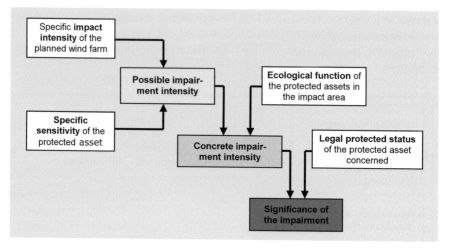

Fig. 3. Influence factors in the assessment of impairments

How the influence factors for assessment shown will be manifested concretely in an actual case depends on the correlations of impacts to be assessed. For example, while in case of habitat loss of sea birds, the area size of the wind farm and turbine density are the primary determining factors, for marine mammals, the main factors determining the impact intensity of offshore wind farms are noise emissions caused by ramming of the foundations. In these examples, the specific sensitivity of the protected assets in relation to the turbines is described by the disturbance sensitivity of sea birds, or by the hearing sensitivity of marine mammals, respectively. Thus, depending on which of the impact correlations identified as relevant for the decision-making process is to be considered, the specific impact factors in detail may be very different.

Moreover, the concrete impairment intensity is also dependent on the technical significance of the protected assets in the impact area of the planned facility (e.g. the ecological function of the local population). Which impairment intensity will ultimately be viewed as so considerable that it meets the criteria of endangerment of the marine environment will then additional also depend on the legal significance of that population concerned (e.g. legally defined protected status).

Due to the legally somewhat imprecise specifications of grounds for rejection formulated in § 3, SeeAnlV, it is the duty of the authorisation agency to specify these more concretely. As a rule, not only explicitly formulated legal standards, but also scientifically well-founded, technical evaluation criteria and standards must provide the basis, to permit the imprecise legal stipulations to be concretised and permit an assessment on a legally secure foundation.

18.3.6 Threshold of "Endangerment of the Marine Environment"

Due to the legally stipulated authorisation prerequisites in the Marine Facilities Ordinance, the authorisation authority has no planning-related or other decision-making discretion (Langenheld et al. 2004). It merely has the duty of checking whether the particular authorisation prerequisites are present or not, or whether they might be achieved through the imposition of conditions. Accordingly, an Environmental Impact Assessment suited for enabling a decision must address the question as to whether the impairments connected with the construction, operation or installation of an offshore wind farm are to be assessed as so considerable that the authorisation of the project must be rejected for reasons of endangerment to the marine environment. For a decision on the authorisation of a project, concrete defined thresholds of significance would ideally be required, the surpassing of which would constitute endangerment of the marine environment under the Marine Facilities Ordinance and hence cause the rejection of the project. In practice, however, these thresholds are very difficult to define.

In principle, these endangerment thresholds must be distinguished from thresholds in the scientific sense (e.g. critical loads). The latter describe effect thresholds starting at which changes in the species or protected assets concerned can be observed, ascertained or assumed, or effects on the population is to be expected (Lambrecht et al. 2004). On the other hand, the endangerment thresholds in the legal sense define the socially tolerated degree of impairment, the transgression of which will trigger legal consequences. However, if a scientific threshold describes the point, for example, at which a habitat loss for sea birds would be so great as to endanger the preservation of the population, or threaten the extinction of that population, and the legally defined goal at the same time stipulates the maintenance of that population, this scientific threshold would at the same time also define the threshold as of which habitat loss would mean endangerment of the marine environment.

For the definition of significance thresholds with regard to the determination of the authorisation of a project, the question as to how the impairments concerned are exactly formulated is first of all of central importance, i.e. precisely which change is to be evaluated for the decision on the authorisation of the project.

In the case of food searching sea birds, for example, the threshold to be defined may be oriented towards the size of the entire population of a species, or directly towards the size of the habitat (cf. Fig. 4). At the same time, for example, the choice of a smaller segment of the population of a species (e.g. the regional population) could for example increase the extent of the impairment, while the decision in favour of a larger reference standard (e.g. the bio-geographical population) would lower the significance threshold.

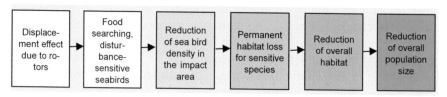

Fig. 4. Cause-effect chain, using the example of impairment of food searching sea birds

Depending on where the threshold starts, the question as to which impact factors are substantial, and with the aid of which indicators and criteria these factors can be ascertained and evaluated at reasonable expense, is to be analysed. Clearly, the availability of qualitative and quantitative information and data will be decisive in determining the degree of distinction for the evaluation models and for the formulation of inadmissibility thresholds.

In the present research project, various approaches regarding the construction and operation of offshore wind farms and the correlations of impacts identified as relevant to the decision-making process have been discussed. However, the determination of threshold values will only be possible if a basic consensus is reached as to the point in the cause-effect chains at which protected asset specific thresholds of significance can be defined. This can occur only in the context of a broad discussion by scientists, administrators involved in the authorisation process, and societal stakeholders.

18.4   Demands upon Data Acquisition in the Context of Environmental Impact Studies

The Environmental Impact Studies (EIS) to be drafted in the context of the Environmental Impact Assessments (EIA) must primarily provide the following:

- A technical and factual basis for the formation of opinion by the various groups involved in the process;
- a basis for a sound authorisation decision-making process, with consideration for the grounds for rejection under § 3 SeeAnlV;
- a basis for an optimisation of projects from the environmental point of view.

On the other hand, objects of investigation which are insignificant for the decision-making process, which can from the outset be assumed to be irrelevant in terms of grounds for rejection, can be neglected in the Environmental Impact Study, if this is confirmed in the scoping process. This includes particularly the objects of investigation human beings, growth on structures, macro-phytobenthos, seabed/sediment structure, air, climate, and material assets and cultural heritage.

For the prognosis and assessment of the effects of offshore wind farms according to the above mentioned principle, e.g. the following information may be required (Table 4, example: marine mammals).

18.5   Results

The accompanying ecological research has produced a large variety of results and findings about occurrence, distribution, density and habitat use of the animal species living in the North and Baltic Seas, and also on the possible consequences of offshore wind farms. It has, however, rapidly become clear that the scientific research results provided do not immediately flow into the required environmental assessment and examination instruments, nor are they transformed into the knowledge required for decision-making processes. 'Translation work' is needed here, from the perspective of a legally standardised decision-making process, in close consultation with natural scientists.

Table 4. Data and information standard for the authorisation of offshore wind farms, using the example of marine mammals

Marine mammals: Damage and/or displacement due to noise emissions when ramming

Information needed for population identification and assessment:	- Harbour porpoise density - Existence of mother-calf groups - Habitat use and spatial distribution - Seasonality of occurrence
Information needed for a prognosis of effects:	**Site-Specific Factors:** - Location of wind farm - Impact area of construction-related acoustic emissions - Type of construction measures (drilling, pile-driving etc.) - Duration and time of construction existing burden, cumulative effect **Non-Site-Specific Factors:** - General sensitivity of marine mammals (avoidance behaviour) - Species-specific sensitivity (hearing) - Life phase (mating, calving, rearing)
Information needed for an assessment of the significance of the effects:	- Ecological function of the area (feeding, raising of young, passage area) - Natural population dynamics - Conservation significance (rare, endangered) - Legal significance - Threshold of intolerable impairment (conservation and/or socially defined targets)

In the context of the present research project, requirements for the contents and procedures of Environmental Impact Studies have been derived from the legal stipulations for the authorisation of offshore wind farms in the German EEZ, as well as from the state of scientific knowledge and discussion on impairment of the marine environment. With the developed discussion platform for the assessment of impairment on the marine environment and bird migration, a base of knowledge which, based on the state of ecological research, develops the steps for assessment required for the decision-making process for the approval of offshore wind farms. It makes suggestions for evaluation criteria and discusses them.

An Environmental Impact Assessment must be decision oriented. That implies that the developer is obliged to submit only such documents on environmental effects which might be relevant to the decision-making process in view of the Marine Facilities Ordinance and the specification standards provided in § 3 therein. Therefore, only such effects which could cause rejection of the project are to be recorded there.

For the assessment of the significance of the effects, or the question as to whether a project constitutes endangerment of the marine environment, it is mandatory that the assessment standards defined in the Marine Facilities Ordinance be applied on the individual protected assets of the marine environment. This actually means in the case of habitat loss of sea birds, for example, that in terms of "endangerment of the marine environment", it is necessary to define how great habitat loss may be at a maximum, before a rejection of approval of an application must be issued. Once the correlations of impacts have been identified which, under the present level of knowledge could constitute endangerment of the marine environment, those points of approaches for thresholds of significance must be identified by means of which transgression can be ascertained. Ultimately, where such a threshold of the inadmissibility is to lie must become a matter of convention. The process of such convention formulation has not yet been initiated; however, the results of the present project constitute a sound preparatory step for it. It is decisive that such a process of discussion and convention formulation not stop at national borders, but be conducted internationally among experts and authorities.

The great variety of new scientific knowledge, be it the results of the German ecological accompanying research programme or the monitoring results from the two large offshore demonstration wind farms in Denmark, will help to better forecast the possible effects of the construction and operation of offshore wind farms. It will therefore be increasingly easier in future to cope with these necessary processes of discussion and to take into account the interests of the marine environment in the Environmental Impact Assessment more adequately.

References

Bernotat D (2003) FHH-Verträglichkeitsprüfung – Fachliche Anforderungen an die Prüfungen nach § 34 und § 35 BNatSchG. In: UVP-report, Sonderheft zum UVP-Kongress 2002, pp 17-26

BSH (Federal Maritime and Hydrographic Agency) (2003) Standards for Environmental Impact Assessment of Offshore Wind Turbines in the Marine Environment

BSH (Federal Maritime and Hydrographic Agency) (2004) Genehmigungsbescheid vom 23.08.2004 zum Offshore Windenergiepark "Sandbank24". Aktenzeichen: 8086.01/Sandbank24/04 Z11

BSH (Federal Maritime and Hydrographic Agency) (2004a) Ablehnungsbescheid vom 20.12.2004 zum Offshore Windpark "Pommersche Bucht"

Köppel J, Langenheld A, Peters W, Wende W, Finger A, Köller J, Sommer S (2003) Ökologische Begleitforschung zur Windenergienutzung im Offshore-Bereich der Nord- und Ostsee: Teilbereich "Instrumente des Umwelt- und Naturschutzes": Strategische Umweltprüfung, Umweltverträglichkeitsprüfung und FFH-Verträglichkeitsprüfung. (UVP-Leitfaden, FFH-VP-Leitfaden, Diskussionsplattform). Research under the future investment programme of the Federal Government, for the Federal Ministry for the Environment, Conservation and Nuclear Safety – FKZ 0327531

Köppel J, Peters W, Wende, W (2004) Eingriffsregelung, Umweltverträglichkeitsprüfung, FFH-Verträglichkeitsprüfung. Verlag Eugen Ulmer Stuttgart

Lambrecht H, Trautner J, Kaule G, Gassner E (2004) Ermittlung von Erheblichen Beeinträchtigungen im Rahmen der FFH-Verträglichkeitsprüfung. FU-Vorhaben im Rahmen des Umweltforschungsplanes des Bundesministeriums für Umwelt, Naturschutz und Reaktorsicherheit im Auftrag des Bundesamtes für Naturschutz – FKZ 801 82 130 [unter Mitarbeit von M. Rahde u.a.]. – Endbericht: 316S. –Hannover, Filderstadt, Stuttgart, Bonn

Langenheld A, Finger A, Köppel J, Kraetzschmer D, Peters W, Wende W (2004) Methoden zur Beurteilung von Eingriffen in Ökosysteme – am Beispiel der Umweltwirkungen von Offshore-Windparks. Handbuch der Umweltwissenschaften – 12. Erg. Lfg. 6/04

Morkel L (2005) Bewertung in der Umweltverträglichkeitsprüfung von Offshore Windparks – Fallstudienanalyse als Beitrag zur Qualitätssicherung. Diplomarbeit an der TU Berlin, Fachgebiet Landschaftsplanung, insbesondere Landschaftspflegerische Begleitplanung und Umweltverträglichkeitsprüfung

International
Ecological Research

19 European Review of Environmental Research on Offshore Wind Energy

Elke Bruns, Ines Steinhauer

19.1 Introduction

In recent years, many countries have started to develop a considerable potential of wind power at sea to increase the use of renewable energies. At present, long-term planning in Europe provides for more than 50,000 megawatts of offshore wind energy capacity (EWEA 2004). The majority of offshore plans have been drawn up by Denmark, United Kingdom, Germany, Sweden, Netherlands, Ireland, and Belgium.[1] Other than in Germany, most offshore developments are planned in near-shore areas.

In such other countries as Poland, Spain and France, the development has only just started (cf. Greenpeace 2005). Some non-European countries, including the United States, Canada and China, have also put forward offshore wind farm plans. However, outside of Europe the development of offshore wind is not yet far advanced, and has so far been restricted to the planning of individual projects.

In north-western Europe, the first offshore wind farms are already in operation, most of them small near-shore projects (cf. COD 2005a, 8; Greenpeace 2005, 23 *et seq.*), but including, too, six larger projects which can be considered as representative for future wind farms at sea: Horns Rev and Nysted in Denmark, Arklow Bank in Ireland, and North Hoyle, Scroby Sands and Kentish Flats in the United Kingdom.

This chapter will provide an overview of current research activities on environmental issues of offshore wind energy in Europe. The focus is on countries which have implemented research activities comprising more comprehensive and thorough investigations than usually required by European Directives or national licensing stipulations for approval of individual projects.

---

[1] The countries mentioned are members of the EU project Concerted Action for Offshore Wind Energy Deployment (COD). For more information, see: http://www.offshorewindenergy.org/index_cod.php

19.2   Environmental Research on Offshore Wind Energy

In connection with offshore wind energy development, a number of different types of environmental studies are currently being conducted in the countries named. Overall, a distinction can be made between comprehensive research projects on marine environment in general (so called generic research projects) on the one hand, and project-related environmental assessment and monitoring studies on the other (cf. COD 2005b).

Generic research projects are as a rule not connected to specific wind farm projects, and cover a wide range of marine environmental topics relevant to impact assessment procedures. They include baseline surveys for selected environmental features (e.g. occurrence and distribution of sea birds). These data collections are not exclusively linked to offshore facility assessment, but also apply for the delineation of national marine protection areas in some countries.

Other generic research topics deal with methods and techniques for measuring and screening pressure factors exerted by offshore wind facilities. The investigation of occurrence, spread and intensity of specific environmental effects on the marine environment (e.g. submarine noise calculations) is an essential precondition for the performance of prognosis and assessment procedures.

To a lesser extent, methodological aspects of environmental risk assessment and planning procedures are also addressed.

Other research projects linked with specific offshore wind farms provide project-related environmental assessment and monitoring studies. Aiming at the monitoring of specific effects on components of the marine environment, they comprise baseline surveys conducted prior to construction in the context of the Environmental Impact Assessment (EIA) and – depending on the respective licensing conditions – monitoring studies during construction and operation. To date, the most extensive effect studies have been provided by the Danish monitoring programme conducted at the pilot wind farms at Horns Rev and Nysted.

In the following, a review of the environmental research on offshore wind energy is given for the relevant countries:

19.2.1 Denmark

After realisation of several small near-shore projects, Denmark decided in the late '90s to establish large-scale demonstration projects for offshore wind energy. In this context an extensive environmental programme has been initiated to investigate the effects of offshore wind farms. It was implemented at two projects:

- Horn Rev, with 80 turbines in the North Sea, and
- Nysted, with 72 turbines in the Baltic Sea.

The programme took departure in the baseline studies initiated in 1999 and runs through the end of 2006. The environmental programme is financed as a public service obligation (PSO). For this purpose, some €11 million have been allocated for the period 2001 through 2006 (Nielsen 2005).

The environmental studies carried out at Horns Rev and Nysted comprise baseline surveys covering two years, monitoring during construction, and monitoring for approximately two years of operation. The surveys particularly focus on sea and migrating birds and marine mammals; fish, benthic communities, hydrology, geomorphology and landscape/scenery are also being investigated. Such aspects as acoustic emissions, electromagnetism and socio-economic issues are also being addressed. The studies at the demonstration projects include for example:

- Monitoring of the number and distribution of staging, moulting and wintering birds in the wind farm areas;
- visual and radar observations to investigate changes in bird migration routes;
- investigations on the collision risks for birds, e.g. using TADS (Thermal Animal Detection System);
- monitoring of harbour porpoises by visual surveys and PODs (acoustic porpoise detectors);
- aerial surveys, satellite tracking and video monitoring of seals;
- monitoring of fish communities, e.g. sandeel investigations and studies on the effects of electromagnetic fields of cables on fish migration;
- hard bottom substrate monitoring;
- infauna monitoring;
- modelling of morphological changes;
- sociological investigations of the acceptance of wind farms by local communities;

- noise measurements.

The monitoring programme is run by the Environmental Group, consisting of the Danish Energy Authority (DEA)[2], the Danish Forest and Nature Agency, Elsam and Energi E2. The studies are carried out by commissioned research institutes and consultants. The work is accompanied and evaluated by an International Advisory Panel of Experts on Marine Ecology (IAPEME).

The advanced nature of the environmental studies of the Danish demonstration programme is unique in the world to date. Their results and experiences therefore provide valuable knowledge for all countries dealing with offshore wind development.

Environmental reports including the preliminary results of the monitoring studies are available on the Horns Rev[3] and Nysted[4] websites. Intermediate results have also been presented at several conferences, particularly at a special conference held in Billund 2004.[5] The conclusions from the environmental programme are to be published at a final conference at the end of November 2006.

Recently, two research projects conducted at Danish offshore wind farms started within the framework of the Joint Declaration on environmental research on offshore wind energy between Denmark and Germany. This will strengthen the European exchange of information and data on environmental impacts.

19.2.2 United Kingdom

The main bodies involved in the development and regulation of offshore wind energy in the UK are the Crown Estate, the Department of Trade and Industry (DTI), the Department for Environment, Food and Rural Affairs (DEFRA) and the Department for Transport (DfT). To date, three offshore wind farms are operational in the UK, with one more under construction.

---

[2] The Danish Energy Authority is the competent authority for the consent and approval of offshore wind farm projects in Denmark.

[3] see: http://www.hornsrev.dk/Engelsk/default_ie.htm

[4] see: http://uk.nystedhavmoellepark.dk/frames.asp

[5] Conference on Offshore Wind Farms and the Environment, 21-22 September 2004, Billund, Denmark (organised by the Danish Energy Authority, the Danish Forest and Nature Agency, Elsam Engineering and Energi E2).

As part of the first round of approvals for offshore wind farms in 2001, the Crown Estate[6] has established a trust fund which contains the interest from refundable deposits paid by developers. This fund is used to finance generic environmental studies. The programme is named COWRIE (Collaborative Offshore Wind Research into the Environment). Its steering committee has identified four priority areas for generic research:

- Potential effects of electromagnetic fields on fish,
- baseline methodologies for aerial and boat based bird surveys,
- displacement of birds (especially the common scoter) from benthic feeding areas, and
- potential effects of underwater noise and vibration on marine mammals.

Commissioned research institutes and consultants are working on these projects. Reports are posted on the COWRIE website.[7]

COWRIE has been extended for the second round of approvals for offshore wind farms. The Crown Estate is placing the non-refundable option fees paid by the Round Two developers into a separate trust fund for the purposes of generic environmental research, data management and education. Because of the size of the COWRIE fund it was decided to register it as a company (COWRIE Ltd) in 2005.

In addition to COWRIE, the DEFRA and the DTI are also funding on-going research projects on offshore wind energy and the environment. For example, three projects are currently being undertaken by the Centre for Environment, Fisheries and Aquaculture Science (CEFAS)[8]:

- Assessment of the significance of changes to the inshore wave regime as a consequence of an offshore wind array;
- development of generic guidance for sediment transport monitoring programmes in response to construction of offshore wind farms;
- investigation of the potential range of socio-economic impacts on the fishing industry from offshore developments.

To gain more knowledge on the potential impacts of offshore wind farms the British Government recently launched an extensive programme of environmental research. To co-ordinate the work a Research Advisory

---

[6] The Crown Estate owns the seabed out to the limit of territorial waters, but also has management responsibilities for projects outside these limits in the Renewable Energy Zone (REZ). It issues leases for wind farm sites at tender sessions.

[7] see: http://www.offshorewindfarms.co.uk

[8] CEFAS is an executive agency of the Department for Environment, Food and Rural Affairs (DEFRA).
For more information, see: http://www.cefas.co.uk/renewables/

Group (RAG)[9] was latterly created by the Department of Trade and Industry. The initial programme budget, allocated by DTI, amounts to £ 2.5 million (about € 3.6 million). A number of projects are currently underway, for example:

- Aerial surveys of water birds in strategic wind farm areas;
- further developing and enhancing the capacity of surveyors for collecting acceptable quality of data on seabird distribution in UK waters;
- production of methodology for assessing the marine navigation safety risk of offshore wind farms;
- guidance for offshore wind farm developers on Seascape Impact Assessment;
- a study to assess fishing activities that may be carried out in and around wind farms.

To plan further research activities, the RAG has compiled a list with research subjects under consideration. The themes include birds, marine mammals, the seascape, fish, shellfish and benthos, seabed and coastal processes, and navigation.

The above mentioned generic research projects are entirely independent of the requirements incumbent on developers to undertake site investigations to inform the Environmental Impact Assessments, or site monitoring requirements which will be specified in conditions attached to licences. Detailed conditions apply to all construction licences for offshore wind farms in the UK regarding the implementation of environmental baseline and monitoring studies (cf. CEFAS 2004, DTI 2004). Generally, depending on the subject, between one and two years of pre-construction studies, construction monitoring and about three years of post-construction monitoring are required by authorities. Details vary between wind farm sites, but common issues include monitoring of sedimentary and hydrological processes, benthic ecology, fish, birds, electromagnetic fields and submarine noise/vibration and its effects on marine mammals.

19.2.3 Netherlands

In the Netherlands, two small wind farms in sheltered waters were built during the mid '90s. In connection with the Dutch government's target to expand offshore wind energy, permits were issued for two new projects: the Near Shore Wind Farm (NSW) and the Q7 wind farm.

---

[9]   For more information, see:
    http://www.dti.gov.uk/renewables/renew_2.1.3.7.htm

The Near Shore Wind Farm is a demonstration project of 100 MW, supported by the government with a subsidy of € 27 million from the CO_2 reduction policy. The project is primarily intended to gain knowledge and experience for future development of large-scale wind farms further out in the North Sea. An extensive Monitoring and Evaluation Programme, the MEP-NSW, has been attached to the Near Shore Wind Farm project.[10]

The structure of the MEP-NSW is divided into two groups: a) Technology and Economy; and b) Nature, Environment and Use Functions (Ministry of Economic Affairs et al. 2001). The latter focuses on the following issues:

- Birds: flight patterns, occurrence, intensity, season, day/night in relation to estimated collision risk;
- birds: disturbance of habitat/forage area;
- birds: barrier effect;
- valuation of the landscape and habituation to the wind farm;
- impact of underwater noise on fish and marine mammals;
- variation and densities of underwater life and the function as a refuge;
- consequences for North Sea users, particularly commercial fishers;
- risks to shipping and consequential damage;
- consequences for mining of minerals and raw materials;
- morphological changes.

Environmental baseline surveys were conducted in 2003/4 by the MEP-NSW project organisation, which is managed by Senter-Novem, in cooperation with the National Institute for Coastal and Marine Management (RIKZ). The baseline studies form the basis for the monitoring of effects during and after construction. The procurement of reference studies was divided into six lots (sections): benthic fauna, demersal fish, pelagic fish, marine mammals, marine birds, and non-marine migratory birds/ flight patterns of migrating birds (both marine and non-marine). Final reports on the baseline studies (available on the MEP-NSW website) contain a description of applied research methods and results of baseline investigation, such as data surveys of investigated stock. Some of the NSW baseline studies were conducted in combination with the baseline studies for the Q7 wind farm. Currently, the Near Shore Wind Farm is under construction.

Concerning the funding of the investigations, the distinction between "necessary" and "desirable" learning goals is the basis for sharing responsibilities and costs: The government was responsible for the baseline studies, and is responsible for a limited number of necessary learning objec-

---

[10] For more information, see: http://www.mep-nsw.nl/

tives. The operator of the Near Shore Wind Farm (NoordzeeWind) is responsible for the rest of the necessary objectives and for the facultative components of the programme. The MEP-NSW project organisation monitors the environmental data collection.

For the other Dutch wind farm projects, monitoring of environmental effects will be the developer's obligation, specified in the permit conditions.

19.2.4 Sweden

In Sweden, three small offshore wind farms have been built to date, and permits have been issued for another four projects: Lillgrund, Utgrunden (II), Klasården and Karlskrona.

In 2003 the Swedish government requested the Swedish Energy Agency (STEM) to initiate a pilot project on wind energy offshore and in the mountain areas. For this project the government allocated SEK 350 million (about € 37 million), to be used over a five years period ending in 2007. In 2003, STEM invited interested companies and scientists to apply for project funding. The proposals submitted included applications both for investment subsidies and for research projects. STEM decided to separate these different types of applications, and asked the Swedish Environmental Protection Agency (Swedish EPA) to manage a research and knowledge programme called **Vindval**,[11] with the objective of initiating applied research, synthesis and analysis on the environmental effects of wind power plants. For this, SEK 35 million (about € 3.7 million) was allocated.

The Vindval programme is to run from 2005 to 2009. Investigations will be conducted around the two wind parks at Lillgrund and Utgrunden, and also outside of these locations. The two main focuses of the Vindval research programme are on environmental effects of offshore wind power plants, and studies of human attitudes towards wind farms. The environmental effect studies will address:

- Fish,
- marine invertebrates,
- marine mammals (especially Baltic harbour seals),
- hydrography,
- migrating bats, and
- wintering seabirds.

---

[11] For more information, see: http://www.naturvardsverket.se

Surveys will be undertaken before, during and two to three years after wind farm construction. A number of baseline studies have already begun.

Another project involving offshore wind farms is the inventory of offshore banks (Grip 2005). The Swedish Government has commissioned the Swedish Environmental Protection Agency (EPA) to make an inventory of marine species and habitats in twenty offshore banks where interest for wind energy exploitation has been expressed. In the initial stage of the work, the Swedish EPA has selected four of these marine banks that are considered of high protection value, and which are to be excluded from exploitation. The inventory programme started in 2003, and is to report to the government in late 2005 and early 2006.

In general, all offshore wind farm developers applying for a permit are required to carry out studies to assess the environmental impact of the project. Moreover, the permits include conditions on required environmental monitoring studies to be carried out during construction and operation.

19.3 Summary

The currently ongoing research activities contribute to greater confidence and security in the decision making processes. Moreover, research programmes benefit from the experience gained in ongoing Environmental Impact Statements.

Apparently, the countries reviewed have – according to their national deployment strategies – chosen different research approaches to investigate the effects of offshore facilities on the marine environment.

- **Denmark** provides the most valuable information on the actual effects on the marine environment in the North and Baltic Seas by monitoring the effects of the existing demonstration wind farms Horns Rev and Nysted.
- In the **United Kingdom**, further basic research focuses on the areas designed as suitable for offshore wind exploitation. Due to national regulations, there is a stronger focus on also investigating the effects of offshore wind deployment on such other uses as navigation.
- In the **Netherlands**, baseline studies at the Near Shore Wind Farm site have recently been initiated as part of the Monitoring and Evaluation Programme (MEP-NSW). The effects on a wide range of environmental components will be monitored in near-shore site conditions, with a distinction made between "necessary" and "desirable" knowledge.
- In **Sweden**, a new approach has now been initiated, involving the assessment of the sensitivity of potential sites prior to project application.

In addition, a research programme focusing on effect studies is coupled to two new pilot projects.

These national research activities have slightly different focuses, as they are being carried out in areas with differing environmental conditions. Thus, the knowledge gained must be interpreted as single case results. The possibilities for generalisation are still limited. The exchange of knowledge and information on environmental information should be systematically continued in future.

References

Centre for Environment, Fisheries and Aquaculture Science (CEFAS) (2004): Guidance Note for Environmental Impact Assessment in Respect of FEPA and CPA Requirements. June 2004. 45 pp

Concerted Action for Offshore Wind Energy Deployment (COD) (2005a): Work Package 3: Legal and Administrative Issues. Final Report. Drafted by Senter-Novem, Utrecht, 47 pp

Concerted Action for Offshore Wind Energy Deployment (COD) (2005b): Work Package 4: Environmental Issues. Final Report. Drafted by Berlin University of Technology, Berlin, 117 pp

Department of Trade and Industry (DTI) (2004): Guidance Notes, Offshore Wind Farm Consents Process. August 2004, London, 27 pp

European Wind Energy Association (EWEA) (2004): Security of Energy Supply: Offshore Wind Can Be the Answer to Europe's Energy Crunch. News Release 27th September 2004. (available: http://www.ewea.org/documents/0927-Offshore%20WD%20FINAL.pdf)

Greenpeace International (2005): Offshore Wind - Implementing a New Powerhouse for Europe. Amsterdam, 164 pp

Grip, Kjell (2005): Considerations of Concerns of Marine Nature Conservation in Planning and Decision-Making Procedures for Offshore Wind Farms in Sweden. In: Technische Universität Berlin; Bundesamt für Naturschutz (2005): International Exchange of Experiences on the Assessment of the Ecological Impacts of Offshore Wind Farms. Proceedings of the International Expert Workshop held at the TU Berlin, Germany, 17-18 March 2005. Preliminary Version of 1 September 2005

Ministry of Economic Affairs; Ministry of Housing, Spatial Planning and the Environment; Ministry of Transport, Public Works and Water Management; Ministry of Agriculture, Nature Management and Fisheries (2001): Near Shore Wind Farm Monitoring and Evaluation Programme (NSW-MEP). October 2001. 47 pp

Nielsen, Steffen (2005): Consideration of Concerns of Marine Nature Conservation in Planning and Decision-Making Procedures in Denmark. Experiences from Horns Rev and Nysted Offshore Wind Farm Demonstration Projects. In:

Technische Universität Berlin; Bundesamt für Naturschutz (2005): International Exchange of Experiences on the Assessment of the Ecological Impacts of Offshore Wind Farms. Proceedings of the International Expert Workshop held at the TU Berlin, Germany, 17-18 March 2005. Preliminary Version of 1 September 2005

Special thanks to Adrian Judd (UK), Walter van den Wittenboer (NL), Kjell Grip and Anders Björck (S) for reviewing the national information on environmental research programmes.

Conclusion and Perspective

20 Conclusion and Perspective

Julia Köller, Johann Köppel, Wolfgang Peters

Unquestionably, wind energy is the vanguard of the expansion of renewable energies in Germany, and just as unquestionably, the future of wind energy is at sea. According to the Federal Ministry for the Environment, Nature Conservation and Nuclear Safety (BMU) the utilisation of offshore wind power holds the greatest potential among renewable energy sources in the medium term for achieving Germany's climate protection goals and its goals for the expansion of the share of renewable energies in the country's electric power supply. At the same time, an evasion to the open sea provides the possibility to avoid increasing conflicts in the settlement of land-based wind power facilities.

However, the use of offshore wind energy is certainly not without ecological risks which pose particular challenges to the planning and design of wind farms. On the one hand, the prospect of expansion provides promises of the realisation of sustainability and climate-protection goals. Wind-power facilities produce clean electric power without the emission of climate-threatening gases. Furthermore, the production of wind energy will contribute to reducing the dependence on fossil fuels and energy imports. On the other hand, however, this form of energy production also carries the potential for negative impacts on the marine environment.

In addition to the compliance with the principle of precaution through the gradual expansion of offshore wind energy, the goal of ecological accompanying research is to eliminate existing uncertainties and knowledge gaps and to ensure an energy production process which is environmentally appropriate and compatible with the conservation of nature.

20.1 Feared Effects

In addition to the effects of particular facilities on areas of marine life and activity, the cumulative effects of the entirety of offshore wind farms off the German coasts are increasingly becoming the focus of attention for possible impacts. Another key issue under discussion, as it had been in the case of wind power facilities on land, was the potential threat to birds. Ascertaining the possible effects of the use of offshore wind power on marine mammals is another challenge. The habitats of these animals are already impaired by a number of factors. In connection with the marine wind-power utilisation, fears were raised that the additional noise which might be generated would drive away especially the sensitive harbour porpoises from their habitats. Additional impairment potentials were also seen in the shading effect and in the electromagnetic fields around submerged cables. The latter was also discussed with reference to fish. Impairments due to sediment plumes and heating of the seabed by electric power transmission and changes in the species composition of the benthos communities were also considered as potential dangers.

In addition to the lack of knowledge of the possible ecological impacts of offshore wind farms on the marine environment, there was also great uncertainty regarding the ecology of the marine environment. The knowledge of the dissemination, the distribution and the composition of the species of the fauna and of the factors which determine marine life was insufficient to permit a precise description of impacts. A resulting and much more serious problem was that due to these gaps of knowledge and information the issues of the marine environment could be addressed only very generally in the framework of the authorisation processes for offshore wind farms.

20.2 Current Knowledge and Consequences of the Gained Information

Due to the broad increase in information gained in the course of ecological accompanying research the initially high degree of uncertainty regarding possible environmental impacts on the marine ecosystem has been eliminated in many areas.

The research on **marine and resting birds** has produced a detailed picture of the seasonal and geographic distribution of the 35 species living in the German North and Baltic Seas. In order to evaluate the significance of the marine area for seabirds two methods were developed; The Wind Farm Sensitivity Index (WSI) and the application of population referenced threshold values for the evaluation of the possible impact on sea and resting birds are the two major results in this research field. By means of the WSI it is now possible to identify areas for the siting process of offshore facilities which, due to their vulnerability, should be off limits to marine wind power utilisation. Moreover, the WSI is a helpful tool for the assessment of impacts on seabirds in the context of Environment Impact Studies. The results of the research on the distribution of sea and resting birds show that the coastal areas of the south-eastern North Sea are very vulnerable to this type of utilisation. The possible impairments on seabirds caused by offshore wind farms in the German part of the North Sea vary according to the species involved and the specific population size. The survey on red-throated divers and common guillemots, for instance, shows that if the presently planned number of wind farms was realised, large parts of the German Bight would no longer be usable by these two seabird species.

In addition to the research on marine and resting birds, information on **migratory birds** could also be gathered in the framework of the BeoFINO-research project with the aid of various methods; methods including radar, thermal imaging and visual as well as acoustic observations. The research has shown that a combination of these methods is most helpful for ascertaining the complexity of the migration phenomenon of birds over the German North Sea. The results clearly confirm that there are intensive migration periods twice a year – during springtime and in the autumn. During the period of research on Helgoland a total of more than 425 different species, which were crossing the German Bight, could be identified. It was possible to confirm that the intensity and time of migration as well as the flying altitude of the species is greatly depended on the season and the weather conditions. More than half of the species observed flew at a height which makes a collision with a wind turbine at least theoretically possible.

Even if all impacts of offshore wind energy facilities on **marine mammals** have not been completely ascertained yet, the basic knowledge was certainly broadened by the extensive investigations conducted by the MINOS research project on harbour porpoises and seals. For the first time, a comprehensive investigation in the density, the distribution patterns and the habitat utilisation of marine mammals in the German EEZ of the North and Baltic Seas has taken place. It was discovered that harbour porpoises use the German North Sea throughout the whole year, with the density of the harbour porpoises decreasing from west to east. The investigations have shown that neither tides nor time of day make any difference in the presence of the animals. Furthermore, it was possible to consolidate the hitherto uncertain knowledge on the dissemination and density of harbour porpoises in the Baltic Sea, where their densities are generally much lower. Both in the North Sea and in the Baltic Sea the MINOS study could ascertain a seasonal gradient of density occurrences as well as a geographical one. On the one hand, the results show a basically decreasing mean density of harbour porpoise occurrence during the winter months. On the other hand, they've proven an attained density maximum during spring and summer. Here too, the research results now obtained have contributed to distinguishing important marine mammal habitats from less frequently used areas.

The spatial and time referenced activity of seals was also investigated. The goal was to understand the ecology of the animals and their role in the ecosystem in order to be able to assess the environmental effects on them. The results showed that the seals leave the coast and hunt for prey in remote and deeper areas, presumably because of greater food supply. Based on this information it must be assumed that the areas planned for offshore wind farms might overlap with the hunting grounds of the seals. Therefore, a conflict between future wind power utilisation and food-seeking seals cannot be excluded, although the type and scope of the concrete effects are hitherto still unclear.

With reference to possible impairments of the hearing of harbour porpoises, a method was developed in the framework of the research project. With the aid of this method measurements of the hearing could be carried out, both in controlled surroundings and in the wild. By means of this method a complete audiogram of a seal in the air as well as a hearing curve for a harbour porpoise could be produced. Unlike harbour porpoises, seals feature a distinct hearing in the low frequency range. However, the question to which extent the communication and the hearing of marine mammals could be impaired is unfortunately unanswered yet.

The knowledge about the dissemination and density of **fish** in the German North and Baltic Seas is the result of long-time investigations. Regarding the distribution and composition of the species, clear differences have been ascertained between the North and Baltic Seas. While in the German Bight depth and distance from the coast primarily determine the composition of typical fish populations, the composition of the species and the density of occurrence in the Baltic Sea are largely affected by the salinity gradient, which drops from the west towards the east. The first results of possible effects of offshore wind farms on fish have come exclusively from studies undertaken at the two Danish offshore wind farms. The previous assumptions that the construction of offshore wind farms and the resulting local change in habitat structures might have an influence on the fish fauna have not been confirmed by these studies. The investigations during the first two years after construction of all 80 turbines in Horns Rev have not provided any indications of a change of sediment composition in the area of the facilities. An expected effect on sand eels could not be confirmed. A reef effect has been ascertained with reference to species composition and abundance of fish fauna in the vicinity of the foundations. The density of individuals of fish fauna in the area around the artificial reefs is visibly higher. The increased biomass production in combination with its attraction effect on fish is seen as the reason for these observations. The accompanying hydro-acoustic investigations carried out in Horns Rev do not show a significant difference between the construction and reference areas; only weak short-time effects on fish abundance could be observed.

The intensive investigations of the **benthic communities** at the FINO 1 research platform, which have lasted 18 months, do permit tentative conclusions. These conclusions, however, merely concern the first reaction phase of the fauna to the newly introduced structure; long-term aspects could not have been conclusively assessed yet. The epifauna developing in these structures represents a change in the common food chain, since a new quality of the food base is developing, which is especially attractive to larger predators and scavengers. The areas surrounding the FINO 1 research platform demonstrated a clear change, extending even to the soft ground communities. For a single installation such as this research platform, this effect may be considered a spatially limited phenomenon. It is conceivable, though, that the planned wind farms, consisting of hundreds of identical structures, could lead to a considerable cumulative effect. All factors such as the type of underwater structure, the materials used, or the question of whether or not anti-fouling or anti-cratering measures were applied will affect the results. Especially the quantitative and qualitative changes make prognoses regarding larger wind farms very difficult. Initial results of the investigations into possible effects of electromagnetic fields on the marine benthic fauna indicate that even electromagnetic fields of underwater cables do not have any explicit effect on the orientation, the movement or the physiology of the animals tested.

20.3 Further Research and Future Ecological Accompanying Research

Although important information was gained through ecological accompanying research and many open questions were answered, we are still far from having solved every uncertainty and knowledge gap regarding the effects of offshore wind turbines on the marine environment.

Thus, while such phenomena as the behavioural reactions of **harbour porpoises** observed during the construction of the Danish wind farms can unequivocally be identified as escape reactions to expulsion effects, we cannot yet say for sure whether the low frequency underwater sound emitted during the operational phase of the facilities will permanently affect the avoidance behaviour of the harbour porpoises and thus result in habitat loss for these animals, or whether a habituation effect will set in. In this context, the scientists give reason to consider that a habituation to sounds which would potentially drive the animals away does not necessarily mean that no physical damage has been done to the animal.

However, the question remains as to whether or not there are sufficient alternatives and less noisy habitats for the animals if parts of their preferred habitat are ultimately occupied for marine wind power facilities. Moreover, the present state of knowledge is not yet sufficient in order to evaluate the long-term effects on the reproduction and the population development of harbour porpoises.

Should it become apparent that the direct effects of offshore wind farms on **seals** – i.e., impairment of the hearing – are less significant than it had been anticipated, indirect factors such as food availability may have an effect on the population development of seals. Such effects are not necessarily to be assessed as negative, as artificial reefs in the North Sea, which is covered mainly by sandy substrate, form new ecological niches for various benthic organisms and fish. Thus, they are also attractive to seals. Although there is still a lack of knowledge of the origins of possible behavioural changes of seals in the German Bight, the information gained provides insight into various behaviour patterns of these animals, and where they are originated.

However, further investigations into the sensory organs of these animals as well as into the affected environmental factors will be necessary in order to be able to understand how harbour porpoises and seals will react to the changed conditions in their marine habitat.

There is also a need for further research into the ascertainment and assessment of the sensitivity of various **seabird species** to the effects of offshore wind farms in order to permit a more exact quantification of possible habitat losses due to the operation of the facilities. With reference to the authorisation capability of future facilities, an evaluation of the cumulative effects on seabirds – e.g. habitat loss – is a crucial factor. In the course of the evaluation of the habitat loss of seabirds the effects of particular wind parks which are to be authorised as well as the already approved facilities should be considered.

A continuous record of flight movements and bird strike events is necessary for the investigation of the collision risk of **migratory birds** in various weather situations and at different times of the day. On the basis of the present research results on possible disorientation of birds by large-scale wind farms or attracting effects due to the safety lighting, no conclusions can be drawn so far. The barrier effects on the energy budgets of resting and migratory birds caused by the turbines are still unknown as well. They should be investigated more thoroughly in the course of currently running projects. Concerning the **fish fauna,** scientists do not assume that the structural changes in and around the structure area will have any direct effects on the population of most species; long-term small-scale effects are likely to be more complex, though. A future analysis of these effects and their appropriate handling will require a very extensive ecological approach – in terms of a long-term analysis of species composition, of investigation into possible interactions, and of a quantification of the process of the changes in the biological and physical environment of the fish fauna.

There is a further lack of knowledge regarding the effects of offshore wind farms on **benthic habitats and their communities.** The long-term changes in the composition of the species communities can only be ascertained in long-term investigations and with large-scale references that go far beyond the scope of the investigations at the FINO 1 research platform. A possible succession to a more mature community of longer living species will require a much larger time frame as well as a larger spatial scope. During the first observation phase at the FINO 1 research platform a stable situation in the species composition of the benthos communities has not yet been achieved. For an analysis of the continuing development as well as the observation of cumulative effects resulting from a large number of foundations of future wind parks investigations will be necessary to close the existing knowledge gaps.

20.4   International Coordination of Research and Exchange of the Information

In the future, research efforts on the issue of cumulative effects will be increasingly necessary. In view of the expanse of planned wind farms at sea, the knowledge of the effects of single offshore wind turbines on marine habitats will be useful only to a very limited degree. Most scientists argue that the knowledge base of synergy effects and indirect ecological impacts is currently too little to permit a sound decision on the authorisation of an ample number of large wind farms.

Ecological research and the new findings it has achieved are of great importance for the assessment of impacts. They also show, however, that many of the initially expressed fears of possible impairments of marine habitats due to the construction and operation of offshore wind farms could not be confirmed. The ecological accompanying research should be continued permanently in order to be able to take all relevant impacts into account in future authorisation proceedings. The monitoring of existing wind farms should be a central aspect here, with consideration for the strong dynamics of the marine ecosystems and the great variability between different years. The continuation and expansion of the ecological investigations already underway at existing wind farms is also vital for the determination and establishment of appropriate avoidance and minimisation measures. For authorisation decisions, which remain decisions made under conditions of uncertainty, it will be necessary to ensure that the precautionary principle applies, particularly with regard to the still existing knowledge gaps.

Many national and international research projects were initiated in order to close these knowledge gaps. The goal must be to incorporate all results of all these efforts for the authorisation procedures for offshore wind farms. For this purpose, the development of methods for the assessment will have major priority. For all conservationist disciplines connected with the construction and operation of offshore wind farms the task must be to show which habitats will be changed by marine wind power utilisation, and to provide information as to how and to which extent this will happen. However, the assessment of these changes is the responsibility of the society as a whole, not merely of the scientific community. This means, especially in cases of ecologically risky interventions, that society must reach a consensus about acceptable risks. Finally, the question to be answered is: Does a compatibility with the conservation of nature still exist if habitats and communities of species, while not being completely destroyed, will be nonetheless consciously affected in such a way that the original state disappears entirely or that new constellations originate? For, depending on one's point of view, offshore wind farms may be seen as a major impairment of the benthos, or else as a marine biological opportunity because their foundations, acting as artificial reefs, may permit rich habitats with entirely new species to emerge.

Annex

Table 1. Ecological Research Initiated by the German Federal Government in the North and Baltic Seas

Re-search code no.	Period	Recipient/ Implementing entity	Contact	Research topic
0327531	01.11.2001 - 31.08.2003	Technische Universität Berlin - Fakultät VII - Institut für Land-schafts- und Umweltplanung	Prof. Dr. Johann Köppel	Ökologische Begleitforschung zur Windener-gie-Nutzung im Offshore-Bereich der Nord- und Ostsee - Teilbereich: Instrumente des Um-welt- und Naturschutzes - Strategische Um-weltprüfung, Umweltverträglichkeitsprüfung und Flora-Fauna-Habitat-Verträglichkeitsprüfung
0327528A	01.11.2001 - 29.02.2004	Universität Hannover - Curt-Risch-Institut für Dynamik, Schall- und Messtechnik	Dipl.-Ing. W. J. Gerasch	Standardisierte Ermittlung von Immissions-schutzgrenzwerten für die Geräuschentwicklung von Offshore-Windenergieanlagen
0327527	01.01.2002 - 30.04.2004	Technische Universität Hamburg-Harburg - Arbeitsbereich Schiffs-technische Konstruktionen und Berechnungen	Prof. Dr.-Ing Eike Lehmann	Rechnerische Bewertung von Fundamenten von Offshore Windenergieanlagen bei Kollisionen mit Schiffen
0327526	01.11.2001 - 31.12.2004	Stiftung Alfred-Wegener-Institut für Polar- und Meeresforschung (Stiftung AWI)	Dr. Alexander Schröder	Ökologische Begleitforschung zur Windener-gienutzung im Offshore-Bereich auf For-schungsplattformen in der Nord- und Ostsee (BEOFINO)

Table 1 (cont.)

Re-search code no.	Period	Recipient/ Implementing entitiy	Contact	Research topic
0327520	01.01.2002 - 31.03.2004	Landesamt für den Nationalpark Schleswig-Holsteinisches Wattenmeer	Dr. Adolf Kellermann	Verbundvorhaben "Marine Warmblütler in Nord- und Ostsee: Grundlagen zur Bewertung von Windkraftanlagen im Offshorebereich" (MINOS)
0327530	01.02.2002 - 31.07.2003	Schreiber Umweltplanung, Dr. Matthias Schreiber	Dr. rer. nat. Matthias Schreiber	Maßnahmen zur Vermeidung und Verminderung negativer ökologischer Auswirkungen bei der Netzanbindung und -integration von Offshore-Windenergieparks
0327525	01.02.2002 - 30.11.2002	Schreiber Umweltplanung, Dr. Matthias Schreiber	Dr. rer. nat. Matthias Schreiber	Ableitung der fachlichen Kriterien zur Ermittlung von besonderen Schutzgebieten nach Artikel 4 der Vogelschutzrichtlinie und Vorschlagsgebieten nach Artikel 4 der Fauna - Flora - Habitat - Richtlinie
0329928	01.05.2004 - 30.04.2006	Technische Universität Hamburg-Harburg - Arbeitsbereich Schiffstechnische Konstruktionen und Berechnungen	Prof. Dr.-Ing Eike Lehmann	Rechnerische Bewertung des Risikos herabstürzender Gondeln von Offshore-Windenergieanlagen bei der Kollision mit Schiffen
0329945	01.07.2004 - 31.08.2005	Universität Lüneburg - Fachbereich IV Umweltwissenschaften	Prof. Dr. Thomas Schomerus	Strategische Umweltprüfung und strategisches Umweltmonitoring

Table 1 (cont.)

Re-search code no.	Period	Recipient/ Implementing entity	Contact	Research topic
0329946	01.06.2004 - 30.04.2007	Landesamt für den Nationalpark Schleswig-Holsteinisches Wattenmeer	Dr. Klaus Koßmagk-Stephan	Verbundvorhaben MINOS plus: Weiterführende Arbeiten an Seevögeln und Meeressäugern zur Bewertung von Offshore-Windkraftanlagen
0329947	01.07.2004 - 30.06.2006	Universität Hannover - Curt-Risch-Institut für Dynamik, Schall- und Messtechnik	Dipl.-Ing. W.J. Gerasch	Standardverfahren zur Ermittlung und Bewertung der Belastung der Meeresumwelt durch Schallimmissionen von Offshore-WEA
0329948	01.07.2004 - 30.06.2006	Institut für Angewandte Ökologie Forschungsgesellschaft mbH Broderstorf bei Rostock	Prof. Dr. Holmer Sordyl	Artbezogene Erheblichkeitsschwellen von Zugvögeln für das Seegebiet der südwestlichen Ostsee bezüglich der Gefährdung des Vogelzuges im Zusammenhang mit dem Kollisionsrisiko an Windenergieanlagen
0329949	01.06.2004 - 31.05.2006	Technische Universität Berlin - Institut für Landschaftsentwicklung - Fachgebiet Landschaftspflege und Naturschutz	Prof. Dr. Johann Köppel	Berücksichtigung von Auswirkungen auf die Meeresumwelt bei der Zulassung von Windparks in der Ausschließlichen Wirtschaftszone
0329954	01.09.2004 - 31.12.2006	Institut für Angewandte Ökologie Forschungsgesellschaft mbH Broderstorf bei Rostock	Prof. Dr. Holmer Sordyl	Einsatz von Biomarkern für die Erfassung möglicher Wirkungen von elektromagnetischen Feldern (Teil A) und Temperaturen (Teil B) auf marine Organismen (Miesmuscheln und Schlickkrebs) unter Laborbedingungen

Table 1 (cont.)

Re-search code no.	Period	Recipient/ Implementing entitiy	Contact	Research topic
0329955	01.07.2004 - 31.12.2004	Stadt Friedrichstadt	Herrmann Siegfried	Erarbeitung eines fachlichen Konzeptes für ein Info - Center zur Technologie und zu den Technologiefolgen der Offshore - Windenergienutzung
0329946D	01.06.2004 - 30.04.2007	Leibniz-Institut für Meereswissenschaften an der Christian-Albrechts-Universität zu Kiel	Prof. Dr. Dieter Adelung	Verbundvorhaben MINOS plus: Seehunde in (SIS) - Untersuchungen zur räumlichen und zeitlichen Nutzung der Nordsee durch Seehunde im Zusammenhang mit der Entwicklung von Offshore-Windenergieanlagen (WEA) Teilprojekt 6
0329946B	01.06.2004 - 30.04.2007	Christian-Albrechts-Universität zu Kiel - Forschungs- und Technologiezentrum Westküste	Dr. Ursula Siebert	Verbundvorhaben MINOS plus: Weiterführende Untersuchungen zum Einfluss von Offshore-Windenergieanlagen auf marine Warmblütler im Bereich der deutschen Nord- und Ostsee (Teilprojekte 1 - 5)
0329946C	01.06.2004 - 30.04.2007	Deutsches Meeresmuseum Stralsund	Dr. Harald Benke	Verbundvorhaben MINOS plus: Untersuchungen zur Raumnutzung durch Schweinswale in der Nord- und Ostsee mit Hilfe akustischer Methoden (PODs) Teilprojekt 3

Table 1 (cont.)

Re-search code no.	Period	Recipient/ Implementing entitity	Contact	Research topic
0329957	01.09.2004 - 31.08.2007	Institut für Ostseeforschung an der Universität Rostock (IOW)	Prof. Dr. Hans Burchard	Verbundvorhaben QuantAS - Off: Quantifizierung von Wassermassen - Transformationsprozessen in der Arkonasee - Einfluss von Offshore-Windkraftanlagen
0329957A	01.09.2004 - 31.08.2007	Universität Rostock - Fachbereich Maschinenbau und Schiffstechnik - Lehrstuhl für Strömungsmechanik	Prof. Alfred Leder	Verbundvorhaben QuantAs - Off: Quantifizierung von Wassermassen - Transformationsprozessen in der Arkonasee - Einfluss von Offshore - Windkraftanlagen, Teilprojekt: Fließexperimente
0329957B	01.09.2004 - 31.08.2007	Universität Hannover - Fakultät für Bauingenieurwesen und Geodäsie - Institut für Strömungsmechanik und Elektronisches Rechnen im Bauwesen	apl. Prof. Mark Markofsky	Verbundvorhaben QuantAS - Off: Quantifizierung von Wassermassen - Transformationsprozessen in der Arkonasee - Einfluss von Offshore Windkraftanlagen, Teilpojekt: Numerische Nah - Feld Modellierung
0329963	01.02.2005 - 31.01.2007	Universität Hamburg, Biozentrum Grindel	Dr. Veit Henning	Untersuchungen über die Kollisionsgefahr von Zugvögeln und die Störwirkung auf Schweinswale in den Offshore-Windenergieanlagen Horns Rev, Nordsee, und Nysted, Ostsee, in Dänemark

Table 1 (cont.)

Re-search code no.	Period	Recipient/Implementing entity	Contact	Research topic
0329974B	01.01.2005 - 31.12.2007	Institut für Ostseeforschung an der Universität Rostock (IOW)	Dr. Falk Pollehne	Langfristige Felduntersuchungen zu möglichen Auswirkungen von Offshore- Windenergieparks in der Ostsee (BeoFINO II)
0329974A	01.01.2005 - 31.12.2007	Stiftung Alfred-Wegener-Institut für Polar- und Meeresforschung (Stiftung AWI)	Dr. Alexander Schröder	Benthosökologische Auswirkungen von Offshore-Windenergieparks in der Nordsee (BeoFINO II)
0329983	01.01.2005 - 31.12.2007	Institut für Vogelforschung "Vogelwarte Helgoland", Inselstation Helgoland	Dr. Ommo Hüppop	Auswirkungen auf den Vogelzug - Begleitforschung im Offshore-Bereich auf Forschungsplattformen in der Nordsee "FINOBIRD"
0329963A	01.02.2005 - 31.01.2007	BioConsult Sh, Dr. Georg Nehls	Dr. Georg Nehls	Untersuchungen über die Kollisionsgefahr von Zugvögeln und die Störwirkung auf Schweinswale in den Offshore-Windenergieanlagen Horns Rev, Nordsee, und Nysted, Ostsee, in Dänemark

Index

Index

Printing: Krips bv, Meppel
Binding: Stürtz, Würzburg